CITAB-CTBUH Awards

# China Best Tall Buildings
*An Overview of 2016 China Skyscrapers*

# 中国最佳高层建筑
2016年度中国摩天大楼总览

主编
Edited by

 世界高层建筑与都市人居学会 (CTBUH)
Council on Tall Buildings and Urban Habitat

 中国高层建筑国际交流委员会 (CITAB)
China International Exchange Committee for Tall Buildings

同济大学出版社
Tongji University Press

© Council on Tall Buildings and Urban Habitat (CTBUH), China International Exchange Committee for Tall Buildings (CITAB) and Tongji University Press (TJUP). All rights reserved. Apart from any fair dealing for the purposes of private study, research, criticism or review as permitted under the Copyright Act, no part of this publication may be reproduced, stored in a retrieval system or transmitted in any form by any means, electronic, mechanical, potocopying, recording or otherwise, without the written permission of the copyright proprietors.

© 2016 世界高层建筑与都市人居学会、中国高层建筑国际交流委员会和同济大学出版社保留所有权利，未经版权所有人书面同意，不得以任何形式，包括但不限于电子或实体对本山版物的任何内容进行复制及转载。

*Trademark notice:* Product or corporate names may be trademarks or registeredtrademarks, and are used only for identification and explanation without intent to infringe.

声明：本书中产品名称或公司名称可能是商标或注册商标，仅作标识解释之用，无意侵权。

图书在版编目（CIP）数据

中国最佳高层建筑：2016年度中国摩天大楼总览．汉英对照／世界高层建筑与都市人居学会，中国高层建筑国际交流委员会主编．—上海：同济大学出版社，2016.4
ISBN 978-7-5608-6292-7

Ⅰ.①中… Ⅱ.①世…②中… Ⅲ.①高层建筑—介绍—中国—汉、英 Ⅳ.①TU97

中国版本图书馆CIP数据核字(2016)第075598号

China Best Tall Buildings-An Overview of 2016 China Skyscrapers
中国最佳高层建筑——2016年度中国摩天大楼总览

Edited by 主　编：世界高层建筑与都市人居学会（CTBUH）
　　　　　　　　中国高层建筑国际交流委员会（CITAB）
Executive Editor in Chief 执行主编：Daniel Safarik
Book Coordination/Review 出版统筹／审校：Steven Henry　Daniel Safarik　杜鹏　孙霞
　　　　　　　　　　　　　　　　刘玉姝　陈永佳　姜文伟　陈雷　沈朝晖
　　　　　　　　　　　　　　　　王里禾　陈晟　赵昕
Layout Design 装帧设计：Marty Carver
Typesetting 装帧制作：李政
Cover Design 封面设计：完颖
Translation Coordination 翻译统筹：译言网（www.yeeyan.org）
Translator 翻　译：洪芸　郑嵩岩　吴晓静　沈椿人　许远方
Executive Editor 责任编辑：胡毅
Executive Proofreader 责任校对：徐春莲

出版发行　同济大学出版社 www.tongjipress.com.cn
　　　　　（上海市四平路1239号　邮编：200092　电话：021-65985622）
经　销　全国各地新华书店、建筑书店、网络书店
印　刷　上海安兴汇东纸业有限公司
开　本　889mm×1 194mm　1/16
印　张　13.75
字　数　440 000
版　次　2016年5月第1版　2016年5月第1次印刷
书　号　ISBN 978-7-5608-6292-7
定　价　180.00元

## 致谢

CITAB 与 CTBUH 谨向参与 2016 年"中国高层建筑奖"评奖项目的参赛者、承接信息与图片提交的工作人员、为本书的出版付出努力的所有人表示感谢。

特别感谢 2016 评审组委会为本次评选所付出的时间与精力。

## Acknowledgments

CITAB and CTBUH would like to thank all the organizations and individuals who submitted their projects for consideration in the 2016 awards program, and who undertook the work of submitting information and imagery to make this publication possible.

We would also like to thank our 2016 Awards Jury for volunteering their time and efforts in deliberating this year's selections.

# 目录 | Contents

Congratulations from the Architectural Society of China | 中国建筑学会贺辞   vi
Congratulations from the Architectural Society of Shanghai, China
上海市建筑学会贺辞   vii
Foreword by Chunhua Song | 序/宋春华   viii
Foreword by Junjie Zhang | 序/张俊杰   x
Foreword by David Malott | 序/大卫·马洛特   xii
Foreword by Antony Wood, | 序/安东尼·伍德   xiv
Introduction | 前言   xvi

## China Best Tall Building | 中国最佳高层建筑奖

**Awards Criteria | 评选标准**   1

### Excellence Award | 优秀奖

Asia Pacific Tower & Jinling Hotel, *Nanjing*
金陵饭店亚太商务楼, 南京   2

Bund SOHO, *Shanghai*
外滩SOHO, 上海   8

Hongkou SOHO, *Shanghai*
虹口SOHO, 上海   14

Wangjing SOHO, *Beijing*
望京SOHO, 北京   20

### Honorable Distinction | 荣誉奖

Fake Hills Linear Tower, *Beihai*
假山大厦, 北海   26

Hua Nan Bank Headquarters, *Taipei*
华南银行总部大楼, 台北   30

Lujiazui Century Financial Plaza, *Shanghai*
陆家嘴世纪金融广场, 上海   34

Nanchang Greenland Central Plaza, *Nanchang*
南昌绿地中心, 南昌   38

Nanchang Greenland Zifeng Tower, *Nanchang*
南昌绿地紫峰大厦, 南昌   42

OLIV, *Hong Kong*
香港OLIV, 香港   46

People's Daily New Headquarters, *Beijing*
人民日报新总部, 北京   50

Taiping Finance Tower, *Shenzhen*
深圳太平金融大厦, 深圳   54

Zhengzhou Greenland Plaza, *Zhengzhou*
郑州绿地中心·千玺广场, 郑州   58

### Nominee | 提名作品

5 Corporate Avenue, *Shanghai*
企业天地中心5号楼, 上海   62

Agile Center, *Guangzhou*
雅居乐中心, 广州   64

Changsha Xinhe North Star Delta Center, *Changsha*
长沙北辰新河三角洲, 长沙   66

Changzhou Modern Media Center, *Changzhou*
常州现代传媒中心, 常州   68

Chongqing Land Group Headquarters, *Chongqing*
重庆地产集团总部, 重庆   70

Forebase Financial Plaza, *Chongqing*
申基金融广场, 重庆   72

Fortune Financial Center, *Beijing*
北京财富金融中心, 北京   74

HVW Headquarters, *Taoyuan*
台湾HVW总部, 桃园   76

J57 SkyTown, *Changsha*
J57天空之城, 长沙   78

Ji'nan Greenland Center, *Jinan*
济南绿地中心, 济南   80

Jing Mian Xin Cheng Tower, *Beijing*
京棉新城大厦, 北京   82

Kingtown International Center, *Nanjing*
南京金奥国际中心, 南京   84

Oriental Blue Ocean International Plaza, *Shanghai*
东方蓝海国际广场, 上海   86

Oriental Financial Center, *Shanghai*
东方汇经中心, 上海   88

R&F Yingkai Square, *Guangzhou*
富力盈凯广场, 广州   90

Shanghai Arch, *Shanghai*
上海金虹桥国际中心, 上海   92

Shenzhen Zhongzhou Holdings Financial Center, *Shenzhen*
深圳中洲控股金融中心, 深圳   94

Sunrise Kempinski Hotel, *Beijing*
日出东方凯宾斯基酒店, 北京   96

Center 66, *Wuxi*
无锡恒隆广场, 无锡   98

Colorful Yunnan • Flower City, *Kunming*
七彩云南花之城, 昆明   98

Corporate Avenue 6, 7 & 8, *Chongqing*
企业天地6,7,8号楼, 重庆   99

Dachong Commercial Center, *Shenzhen*
大涌商务中心, 深圳   99

Ding Sheng BHW Taiwan Central Plaza, *Taichung*
鼎盛BHW台湾中心广场, 台中   100

Evergrande Huazhi Plaza, *Chengdu*
恒大华置广场, 成都   100

Fuzhou Shenglong Financial Center, *Fuzhou*
福州升龙汇金中心, 福州   101

Global Harbor, *Shanghai*
环球港, 上海   101

Grand Hyatt Dalian, *Dalian*
大连君悦酒店, 大连   102

JW Marriott Shenzhen Bao'an, *Shenzhen*
深圳前海华侨城JW万豪酒店, 深圳   102

Mount Parker Residences, *Hong Kong*
西湾台1号, 香港   103

Ningbo Global Shipping Plaza, *Ningbo*
宁波环球航运广场, 宁波   103

Shaoxing Shimao Crowne Plaza, *Shaoxing*
绍兴世茂皇冠假日酒店, 绍兴   104

Shenzhen Xinhe World Office, *Shenzhen*
深圳星河World写字楼, 深圳   104

Studio City, *Macau*
新濠影汇, 澳门   105

The Wave of Science and Technology Park S01, *Ji'nan*
浪潮科技园S01科研楼, 济南   105

Tianjin International Trade Tower 1, 2 & 3, *Tianjin*
天津国际贸易中心1,2,3号楼, 天津   106

WPP Campus, *Shanghai*
达邦协作广场, 上海   106

Wuxi Suning Plaza 1, *Wuxi*
无锡苏宁广场, 无锡   107

Xiamen World Overseas Chinese International Conference Center, *Xiamen*
厦门世侨中心, 厦门   107

## China Tall Building Urban Habitat Award | 中国高层建筑城市人居奖

**Awards Criteria | 评选标准**   108

### Winner | 城市人居奖

Jing An Kerry Center, *Shanghai*
静安嘉里中心, 上海   110

### Honorable Distinction | 荣誉奖

Heart of Lake, *Xiamen*
万科湖心岛, 厦门   116

Shenye Tairan Building, *Shenzhen*
深业泰然大厦, 深圳   120

**Nominee | 提名作品**

| | |
|---|---|
| Huizhou Central Place, *Huizhou* 惠州华贸中心，惠州 | 124 |
| Wuhan Tiandi Site A, *Wuhan* 武汉天地A座，武汉 | 126 |

**China Tall Building Legacy Award | 中国高层建筑成就奖**

| | |
|---|---|
| Awards Criteria | 评选标准 | 128 |

**Winner | 成就奖**

| | |
|---|---|
| White Swan Hotel, *Guangzhou* 白天鹅宾馆，广州 | 130 |
| Hong Kong and Shanghai Bank, *Hong Kong* 汇丰银行，香港 | 131 |
| International Foreign Trade Center, *Shenzhen* 国贸中心，深圳 | 132 |
| East China Electrical Power Distribution Building, *Shanghai* 华东电力调度大楼，上海 | 133 |
| Bank of China Tower, *Hong Kong* 中国银行大厦，香港 | 134 |
| Shanghai Center, *Shanghai* 上海商城，上海 | 135 |
| Shun Hing Square, *Shenzhen* 信兴广场，深圳 | 136 |
| Jin Mao Tower, *Shanghai* 金茂大厦，上海 | 137 |
| Two International Finance Center, *Hong Kong* 国际金融中心二期，香港 | 138 |
| TAIPEI 101, *Taipei* 台北101大楼，台北 | 139 |

**Honorable Distinction | 荣誉奖**

| | |
|---|---|
| China Resources Building, *Hong Kong* 华润大厦，香港 | 140 |
| Kunlun Hotel, *Beijing* 昆仑饭店，北京 | 140 |
| Lippo Center, *Hong Kong* 力宝中心，香港 | 141 |
| Jin Jiang Tower Hotel, *Shanghai* 上海新锦江大酒店，上海 | 141 |
| Jing Guang Center, *Beijing* 京广中心，北京 | 142 |
| Central Plaza, *Hong Kong* 香港中环广场，香港 | 142 |
| T & C Tower, *Kaohsiung* 高雄85大楼，高雄 | 143 |
| Bank of China Tower, *Shanghai* 上海中银大厦，上海 | 143 |
| Tomorrow Square, *Shanghai* 明天广场，上海 | 144 |
| Guangzhou Development Center Building, *Guangzhou* 广州发展中心大厦，广州 | 144 |

**Nominee | 提名作品**

| | |
|---|---|
| The Landmark Gloucester Tower, *Hong Kong* 告罗士打大厦，香港 | 145 |
| Hopewell Center, *Hong Kong* 合和中心，香港 | 145 |
| Shenzhen Development Bank, *Shenzhen* 深圳发展银行大厦，深圳 | 145 |
| CITIC Plaza, *Guangzhou* 广州中信广场，广州 | 146 |
| The Center, *Hong Kong* 香港中环中心，深圳 | 146 |
| Dalian World Trade Center, *Dalian* 大连世贸大厦，大连 | 146 |
| Macau Tower, *Macau* 澳门旅游塔，澳门 | 147 |
| Pudong International Information Port, *Shanghai* 浦东国际信息港，上海 | 147 |
| Guangming Building, *Shanghai* 光明大厦，上海 | 147 |

**China Tall Building Innovation Award | 中国高层建筑创新奖**

| | |
|---|---|
| Awards Criteria | 评选标准 | 148 |

**Winner | 创新奖**

| | |
|---|---|
| Mega-Suspended Curtain Wall (Shanghai Tower, Shanghai) 悬挂式巨型玻璃幕墙（上海中心大厦） | 150 |

**Honorable Distinction | 荣誉奖**

| | |
|---|---|
| Cold-Bending Glass at the Supertall Scale (Nanchang Greenland Central Plaza, Parcel A) 超高层冷弯玻璃（南昌绿地中心A座） | 154 |
| Construction-Phase Internal Force Redistribution (Wuhan Center, Wuhan) 利用施工程序调整超高层结构内力分布的设计方法（武汉中心） | 156 |
| Hybrid Outrigger (Raffles City, Chongqing) 创新组合伸臂系统（重庆来福士广场） | 158 |

**Nominee | 提名作品**

| | |
|---|---|
| Curved Façade Connector (Wuhan Center, Wuhan) 曲面建筑表皮连接构造（武汉中心） | 160 |
| Gravity-Driven Fire-Extinguishing System (Zifeng Tower, Nanjing) 重力消防系统（南京紫峰大厦） | 162 |
| Hangzhou Citizen Center 杭州市民中心 | 163 |

**China Tall Building Construction Award | 中国高层建筑建造奖**

| | |
|---|---|
| Awards Criteria | 评选标准 | 164 |

**Winner | 建造奖**

| | |
|---|---|
| Forum 66, *Shenyang* 沈阳市府恒隆广场，沈阳 | 166 |

**Honorable Distinction | 荣誉奖**

| | |
|---|---|
| Beijing Greenland Center, *Beijing* 北京绿地中心，北京 | 170 |

**Nominee | 提名作品**

| | |
|---|---|
| Changsha Xinhe North Star Delta, *Changsha* 长沙北辰新河三角洲，长沙 | 172 |
| Tianjin Kerry Center, *Tianjin* 天津嘉里中心，天津 | 173 |

**China Tall Building Outstanding Achievement Award 中国高层建筑杰出贡献奖**

| | |
|---|---|
| Awards Criteria | 评选标准 | 174 |

**Winner | 杰出贡献奖**

| | |
|---|---|
| Dasui Wang, *ECADI* | 汪大绥，华东建筑设计研究总院 | 176 |

**Information & Index | 信息与索引**

| | |
|---|---|
| 2016 China Awards Jury | 2016年中国高层建筑奖评审委员会 | 182 |
| Index of Buildings | 建筑索引 | 186 |
| Index of Companies | 企业索引 | 188 |
| Image Credits | 图片版权 | 192 |
| About the CTBUH | 世界高层建筑与都市人居学会 | 194 |
| About the CITAB | 中国高层建筑国际交流委员会 | 195 |
| CTBUH Organizational Structure & Members 世界高层建筑与都市人居学会组织架构和会员 | 196 |
| CITAB Organizational Structure & Members 中国高层建筑国际交流委员会组织架构和会员 | 198 |

# 中国建筑学会贺辞
## Congratulations from the Architectural Society of China

"不畏浮云遮望眼，自缘身在最高层。"

衷心祝贺《中国最佳高层建筑》出版，热切希望中国的高层建筑行业创新领先，勇攀高峰！

中国建筑学会理事长
2016 年 4 月 20 日

*There's no fear that floating clouds may shelter my eyes,*
*Just since now here I am, on mountains, to the top-crest.*

I would like to extend my cordial congratulations to the publication of *China Best Tall Buildings* and I sincerely hope that the architecture industry in China will bring forth new ideas and scale new heights!

**Long Xiu**
President of the Architectural Society of China
April 20th, 2016

# 上海市建筑学会贺辞
## Congratulations from the Architectural Society of Shanghai, China

建筑是上海的第一张名片，上世纪国际饭店雄踞中国建筑第一高度半个世纪，新世纪浦东的上海中心、环球金融中心等一大批享誉中外的高层建筑不断刷新高度，成为中国改革开放在世界的形象。

展望未来，中国高层建筑的建设者们一定会继续奋进，再创辉煌！

上海市建筑学会理事长
2016 年 4 月 21 日

Architecture is the first name card of Shanghai. During the last century, the Park Hotel was once the highest building in China for 50 years; and now, in the new century, world famous buildings, including the Shanghai Tower in Pudong and Shanghai World Financial Center, have dominated the skyline of Shanghai and become the image of China's reform and opening up on the world stage.

Looking into the future, I believe that constructors of high-rise buildings in China will definitely make their best endeavors for greater prosperity!

**Jiaming Cao**
President of the Architectural Society of
Shanghai, China
April 21st, 2016

# 序 | Foreword

**宋春华**，首届"中国高层建筑奖评选"评委会主席，建设部原副部长，中国建筑学会原理事长

**Chunhua Song,** *Chairman of the Jury of the first CITAB-CTBUH China Tall Buildings Awards, Former Deputy Minister of the Ministry of Housing and Urban-Rural Development of the P.R.China, Former Chairman of the Architectural Society of China*

中国高层建筑奖项评选分两个阶段进行，在联合主办单位已经开展的大量前期工作（包括网上申报工作、申报材料的整理汇总）的基础上，先由评委进行第一阶段的远程网上初评工作；之后，在上海举办第二阶段正式评选。经过设计和技术两个专家组的评选，共在4个大类约90个项目中评选出4个中国高层建筑优秀奖，1个城市人居优秀奖，1个创新优秀奖，1个建造优秀奖。评委会还在前期初评的基础上，评出了1个杰出贡献奖和10个成就奖，并评出一批荣誉奖和入围奖。

这次评选活动高效务实，进展顺利。中国高层建筑奖在奖项设置、参评范围、专家组成和专业导向等方面，很有特点，给评委们留下了深刻的印象。

**建筑类型和地域范围的设定有特色**。与以往一些重要的建筑设计评奖活动不同，本次评奖建筑类型只针对已经建成的高层建筑这一细分建筑类型和相关衍生对象，没有其他建筑类型混合参评，这就使评奖目标更为集中，突出高层建筑的特征性，有利于一批优秀高层建筑脱颖而出。评奖范围涵盖整个大中华地区，包括了大陆、香港、澳门和台湾。这样设定的范围有利于加强大陆与港澳台地区的相互交流、相互促进、共同提高，体现了"中国高层建筑奖"这个名称的完整含义。

**奖项类别和覆盖面具有针对性**。奖项类别包括了高层建筑的"建筑奖"、"城市人居奖"、"创新奖"、"建造奖"，还有"杰出贡献奖"和"成就奖"。奖项的设置包含了建筑物、创新技术、建造工艺、人居环境、个人贡献、建筑文化遗产等多维度，体现了全方位、多视角的覆盖面，与高层建筑在大型城市建设和人居环境中的重要影响是相呼应的，并深刻地影响着城市的规划、建设和管理以及城市经济、社会、文化、环境的发展，所设奖项必将引起业界和社会对高层建筑的持续关注，并期待其健康有序地发展。

**评审专家组成具有广泛代表性**。此次评奖活动的评审专家由上海市建筑学会、CITAB和CTBUH（世界高层建筑与都市人居学会）等国内外知名的行业组织共同推荐。由国际、国内相关专家组成的评审专家组体现了广泛的代表性，其中国内的专家包括了中国工程院院士、中国建筑学会和上海市建筑学会领导专家、CITAB主席、著名高校专家教授、知名房地产开发商和业主代表等。境外的专家包括CTBUH主席、CTBUH执行理事长、国际知名设计公司代表等。

**评审活动坚持正确的价值导向**。整个评选活动注重把握好技术导向、学术导向和创新导向，坚持全面评价、综合考量的基本原则，遵循适用、经济、绿色、美观的基本方针。因此，在评选过程中，避免了以往高层建筑评选中容易出现的唯高度论、唯规模论、唯炫技论，更加聚焦设计理念和设计技术创新、更加关注地域文化和区域文脉、更加重视人居环境影响和绿色可持续发展。从评选结果看，获奖项目精彩纷呈，代表着中国近年内已经建成的高层建筑的最新面貌和最新水平。

为总结这次评选活动并促进高层建筑建设活动的交流，主办方拟将获奖成果和其他申报项目付印出版，我对本书的编辑出版表示赞许和祝贺。同时，也想借此机会表达自己三个希望，第一是希望中国高层建筑奖项评选一届接一届办下去，越办越好，越办越有影响力。第二是希望上海市建筑学会、CITAB和CTBUH等行业组织共同携手，在中国建筑学会和各界人士的支持下，继续为中国高层建筑的健康发展努力工作。第三是希望中国高层建筑的同行们，面对城市发展的新趋势和新机遇，一定要认真研究高层建筑的发展规律，并在实践中勇于创新、勇于开拓，在全行业的支持下，在社会各界的关心下，继续为我们新时期城市现代化建设和人居环境的改善作出应有的贡献。

是为序。

The selection process for the China Tall Building Awards has been highly efficient and smooth, and the awarded projects are quite characteristic of China's tall building industry, in terms of the composition and range of projects and the professional acumen invested in them, by which the judges were very impressed. Some key observations:

The range of building type and geographical scope was impressive. Different from many awards for building design conducted in the past, this awards program only targets high-rise buildings and related technologies. Without considering other types of buildings, this awards program is more focused, making it easier for high-rise buildings to stand out and be recognized as works of architecture on their own. The geographical scope covers the whole of Greater China, including Hong Kong, Macau, Taiwan and mainland China. This coverage is not only beneficial for strengthening mutual exchanges, mutual improvement and mutual promotion among mainland China, Hong Kong, Macau and Taiwan regions, but is compatible with the spirit of the China Tall Building Awards program.

The award categories and scope were appropriately specific. The awards included the China Tall Building Excellence Award, Urban Habitat Award, Innovation, and Construction, as well as Legacy Award and Outstanding Achievement Award. These awards have various dimensions, including individual buildings, innovative technology, construction techniques, the quality of the living environment, individual contributions, and architectural heritage. This reflects the truly comprehensive nature of the industry, and the importance of high-rise buildings in urban construction and living environment of large cities. With a profound impact on urban planning, development and construction, as well as development of urban economy, society, culture and environment, these awards will certainly draw the attention of the architecture industry and the whole society to high-rise buildings, and thus facilitate their healthy and steady development.

The jury was well composed. The jurors were recommended by domestically and internationally well-known industry organizations, such as the Architectural Society of Shanghai, the China International Exchange Committee for Tall Buildings (CITAB) and the Council on Tall Buildings and Urban Habitat (CTBUH). This evaluation team was therefore quite representative. It had domestic experts, such as members of the Chinese Academy of Engineering, experts of the Architectural Society of China, leaders of the Architectural Society of Shanghai, and the Chairman of CITAB, distinguished professors and experts from universities, well-known real estate developers and proprietor representatives. Overseas experts included the Chairman and Executive Director of as well as representatives of internationally renowned design companies.

This evaluation process upheld good values. The entire evaluation process emphasized orientations of technology, academics, and innovation, adhered to the basic principles of comprehensive evaluation and examination, and followed basic design principles of "applicable, economical, green, and beautiful." Therefore, it has avoided the temptation of placing too much focus on height, scale, or techniques, which are prone to happen in typical evaluations of high-rise buildings; instead, it focuses more on design philosophies, design technology innovation, regional cultural context, impacts on living environment and sustainable green development. From the selection results, we can see the awarded projects are quite outstanding and brilliant. They represent the latest outcomes and the rising level of sophistication in high-rise buildings in China in recent years.

In order to summarize this evaluation process and fruitful exchanges on the subject of high-rise building construction, the organizers propose to publish these award results. Herein, I extend my true appreciation of and congratulations to the participants in this book. At the same time, I would like to take this opportunity to express my three hopes.

Firstly, I hope the China Tall Building Awards will go on and turn out to be better and better, and more and more influential.

Secondly, I hope industry organizations such as the Architectural Society of Shanghai, CITAB and CTBUH will cooperate with each other to promote the healthy development of high-rise buildings in China, under the support of the Architectural Society of China and people from all walks of life.

My third hope is in the context of new trends and opportunities in urban development. It is that our contemporaries in the high-rise building circle will carefully study the development laws of high-rise buildings; that they will be creative and courageous in practice; and that, with the support of the entire industry and the whole of society, they will contribute more to the urban modernization and the living environment improvement in this new era.

I hope this preface can support this purpose.

# 序 | Foreword

**张俊杰**，中国高层建筑国际交流委员会（CITAB）主席，上海市建筑学会副理事长

**Junjie Zhang**, *Chairman of CITAB, Vice President of Architectural Society of Shanghai*

对于中国高层建筑行业，过去的三年注定是不平凡的三年。

这三年里，中国高层建筑在高度和数量方面再攀高峰。在全球 200 m 高度以上建筑总数首次超过 1 000 座的划时代背景下，中国上海落成了 2015 年全世界最高的竣工建筑、也是截至目前世界第二高建筑。与此相对应，在全球 2015 年竣工数量方面，亚洲地区占比约四分之三，中国地区占比更是超过了二分之一。这三年里，中国政府积极推进中国城市工作的进程，2013 年底召开了"中央城镇化工作会议"，并时隔 37 年，于 2015 年底召开了"中央城市工作会议"，明确提出了要尊重城市发展规律，提高城市工作全局性、系统性、发展持续性、发展宜居性、发展积极性等最为重要的城市工作指导原则。在中国的高层建筑、超大型城市综合体和城市综合功能区蓬勃兴起的同时，城市高速发展中日益凸显和累积的问题与矛盾正受到越来越广泛和深入的关注，因此积极推动中国高层建筑和人居环境的研究探索与创新实践工作已经到了迫在眉睫、时不我待的重要历史时刻。

三年前，2013 年 9 月中国高层建筑行业的有识之士们相聚在一起，"中国高层建筑国际交流委员会"（简称 CITAB）的成立构建起了高层建筑领域中国与国际交流的桥梁。

自成立以来，CITAB 及各专业委员会开展了大量工作，2015 年 CITAB 的主要成员单位组队参加了在美国芝加哥举办的 CTBUH 学术峰会；组织高层建筑领域的国际学术活动和优秀高层建筑的评选活动；收集整理中国高层建筑数据，掌握世界高层建筑发展动态和前沿讯息，为中国高层建筑科学研究提供导向；还将接受政府部门、科研单位、企业等委托的有关高层建筑国际交流工作。

在时代和行业发展的背景下，首届"中国高层建筑奖评选活动"由 CITAB 与 CTBUH 携手联合、共同举办。2015 年 10 月活动开始启动，本次活动受到了国内外业界的广泛关注，组办方收集整理了中国大陆和港澳台地区从上世纪 70 年代末到目前为止主要的高层建筑资料，并收到各类奖项申报材料约 90 项。今年 1 月，首届"中国高层建筑奖项评选活动"经过各位国际国内著名专家评委认真、高效、专业、公正的评审，顺利评出了所有奖项，取得了圆满成功。在此，我要感谢所有奖项申报单位的积极支持，特别要感谢宋春华评委会主席领衔的国际、国内评审专家组的辛勤工作和中国建筑学会、上海市建筑学会的大力支持！我要感谢 CTBUH 自始至终高度的重视和鼎力合作。最后我还要感谢本书编辑团队的卓越工作和同济大学出版社的全力支持！

我们将持续改进和全力推进这项工作，为中国高层建筑行业的健康发展，为中国城市化进程的顺利推进和创造优美的人居环境加倍努力！

The last three years have witnessed an increase in the height and numbers of China's tall buildings, marking an extraordinary period. Even though the number of 200m+ tall buildings around the world topped 1,000 for the first time, Shanghai still held its position by completing the second-highest building in the world in the year 2015, and it's still the world's second highest up to now. Asia accounts for three-quarters of the newly completed tall buildings of the world, and of this figure, China alone represents more than a half. This achievement is largely due to the momentum towards urbanization from the Chinese government in the past three years. At the Urbanization Conference of China's Central Government, held in 2013, and the Conference for Central Cities, held later in 2015, the Chinese government put forth important principles to guide urbanization, such as respecting the law of city development, improving the overall importance, development of resilient systems, continuity, habitability and proactivity. With the booming of China's tall buildings, mega-urban complexes and urban comprehensive areas, the problems and contradictions typical of fast urbanization have aroused wider and deeper concerns. As a result, it is urgent to produce more research on China's tall buildings and improve its living condition.

Three years ago, in September 2013, a group of scholars of vision in China's tall building industry got together and set up CITAB, the China International Exchange Committee for Tall Buildings. This committee later became a bridge between the Chinese and international communities of architecture.

CITAB is a branch of the Institute of Chinese Architects. It aims at enhancing the communication about the planning, design, construction, material facilities and operation of China's tall buildings, both for Chinese and overseas institutions and experts. CITAB has carried out a large amount of work since its establishment three years ago. Highlights include participating in a CTBUH academic summit held in Chicago, USA; organizing a variety of international academic activities and awards programs for tall buildings; collecting data on China's tall buildings; tracking the development of tall buildings worldwide; leading the direction of China's architecture study, and performing international exchange programs mandated by government, academia and enterprises.

Given this background, the first China Tall Building Awards program was jointly launched by CITAB and CTBUH in October 2015. This activity drew wide attention at home and abroad. The organizers collected information about major tall buildings in mainland China, Hong Kong, Macau and Taiwan, built from the 1970s to the present, and received about 90 applications in total.

The results came out in January 2016, after the careful and unbiased reviews of Chinese and foreign experts on the jury. Here, I would like to express my gratitude to all applicants, and the judging panel led by Chairman Song Chunhua for their efficiency, carefulness and justice. My gratitude also goes to the Architectural Society of China and the Architectural Society of Shanghai, which Have given us much support. I would also like to say thank you to CTBUH for its high degree of attention and cooperation. Lastly, I also want to say thank you for the excellent work of the editors of the book and Tongji University Press.

We will continually improve our efforts for the sake of the tall building industry, urbanization and living environment of China, as well as introduce the great achievements of China in this field to the world.

# 序 | Foreword

**大卫·马洛特，**CTBUH 主席，KPF 建筑事务所合伙人

**David Malott,** *Chairman, CTBUH; Principal, Kohn Pedersen Fox Associates*

作为有几十年设计高层建筑经验的国外建筑师，我认为首届中国高层建筑奖是一个独特的机会，让我们看到时光变迁中建筑行业的日新月异。如果说中国摩天大厦的繁荣历史，在一开始是西方经验的输入与施行——这种描述并不完全准确，那么现在，更多的中国企业开始通过创新来建造摩天大楼。事实上，在这个辉煌的时期，建筑开发、设计、施工和运营等各个方面都积累了如此多的经验，而这正是中国的沉淀向世界的展示。

该评奖项目设立的原因之一是拓宽中国——这一超高层建筑的重要战地，与世界其他国家思想上的双边交流，而这个行业每年都愈加全球化。这些都是全球最前沿的力量，随着建造压力的增加，建筑既要保证现代化的生活和工作空间，又要使其光鲜亮丽，体现出主办城市无与伦比的动感与活力。而我们的回应，即在这里展示的大多数建筑，用自己得天独厚的条件完美满足了人们的需要，令人耳目一新。其中许多建筑在采用国内创新尖端技术开发的同时，还结合了中国古老的传统元素，适应了当地的自然与城市环境。

参加评审那天阴雨连绵，我走进温暖而繁华的接待区，看到各行各业的专家时，我想起了过去两件印象深刻的事。我的父亲是驻日美国大使，从小我就深刻意识到跨越大洋的沟通合作对整个世界来说多么重要。我发自内心地相信我们所做的只是尽力去改善超高层建筑的环境和社会效用。

另一件事则是很多人都熟悉的，翻过 iPhone 可以看到背面的题字"加利福尼亚设计，中国生产"。这可能也是中国过去建筑业的映射——大多数由国外建筑师设计，但今时不同往日，我们看到了中国建筑天赋的爆发，中国的建筑师和工程师已经成为设计和灵感讨论中不可或缺的一部分。

如今，现代中国的发展不再仅仅从事生产制造，而是显现出向发明创新、技术化和企业化发展的趋势，这在高层建筑行业中尤为明显。本书中所展示的创举大多是国际项目，但实际上，这些建筑的彻底中国本土化才是最激动人心的。中国与世界经济紧密相连，高层建筑行业即是其中一个"巨大的缩影"，中国的发展是整个世界都需要关注的。由此想到这本书和这一高层建筑评奖项目正是承担了一种大使级的工作，它们以肉眼可见的形式代表了中国在世界舞台上所表现出的希望与信任的姿态，同时也预示着我们携手创造的更"高"、更美好的未来。

As an architect with several decades of experience designing skyscrapers in China from abroad, the inaugural China Tall Building Awards program provides a unique opportunity to reflect on how much has changed over that time. If the story of China's skyscraper boom at first was one of Western expertise being "imported" to, and executed in China – and this was never a completely accurate depiction even then – today, much of the innovation in Chinese skyscrapers is happening in China, by Chinese companies. Indeed, there has been so much experience built up on every aspect of development, design, construction and operation during this incredible period, that it is really a case of China having at least as much accumulated knowledge, if not more so, to show the world.

One of the reasons this Awards program was initiated was to widen the opportunities for a bidirectional exchange of ideas between China, the world's capital of skyscraper construction, and the rest of the world, in an industry that becomes more globalized each year. These are the front lines of that globalizing force, as the pressure to rapidly build both contemporary living and working space and shining icons that proclaim the dynamism of their host cities is greater here than anywhere. The best responses to these conditions, many of which we proudly display here, are those that beautifully mesh the requirements with the unique conditions of their settings refreshingly, many of these responses are expressly Chinese, incorporating ancient proverbs and referring to their natural or urban settings, even as they apply cutting-edge technology, much of which is developed domestically.

As I arrived on a rainy day to a warm and buzzing reception area to participate in the judging, seeing the diversity and level of expertise in the room, I was reminded of two strong impressions from my past. I grew up as a US ambassador's son in Japan, and at a very early age it struck me how important it was to have good communication between people who were working together across oceans to solve problems that affect the whole world. I really believe we're doing nothing less than that with this and other efforts to improve the environmental and social performance of high-rises.

The other impression will be familiar to anyone who has looked at the back of an iPhone. The inscription reads, "designed in California, assembled in China." While that may have been true for buildings in China in the past – they were mostly designed by foreign architects – we are now seeing an explosion of talent in China itself, with Chinese architects and engineers becoming an integral part of the design and creative discussion.

Today, the story of modern China is about so much more than production – it's about invention, technology, and enterprise – and nowhere is this truer than in the tall-building world. The incredible feats of engineering and design insight shown in these pages are international projects, to be sure, but what's most exciting is how definitively Chinese they are. Today, China is so deeply linked into the world economy, and the tall-building industry as a "mega-tall microcosm" thereof, that whatever is happening in China is something the world needs to know about. Think of this book and the program it represents as an ambassadorial work that places hope and trust in this most visible aspect of China's position on the world stage, and as a harbinger of not only a taller, but a brighter future we will build together.

# 序 | Foreword

**安东尼·伍德，**CTBUH 执行理事长；美国伊利诺伊理工大学研究教授；同济大学高层建筑方向客座教授

**Antony Wood,** *Executive Director, Council on Tall Buildings and Urban Habitat; Research Professor of Tall Buildings, Illinois Institute of Technology; Visiting Professor of Tall Buildings, Tongji University, Shanghai*

中国高层建筑奖的登场正当其时。过去15年里，中国以其翻天覆地的增长速度震惊了全世界。最明显的证据就是飞速发展的城市中几乎一夜之间拔地而起的摩天大楼。大量的投资，对城市地标的渴求，领导者对城市建设的重视，使很多人认为中国是"建筑师的天堂"，是新型的城市民族，是自由模式的舞台。甚至在几年前，这种趋势就逐渐显现出来。

如今，中国在高层建筑的落成上依然处于世界领先地位，但是大的环境却发生了一些变化。很多人担忧中国的经济增长趋缓——从"天文"数字回归到正常水平——或许不利于建筑业的发展。除此之外，随着政府放出取缔"标新立异"建筑的消息，另一种焦虑也随之而来，许多人怀疑这种繁荣是否能够持续？

答案或许是肯定的，就像毕业典礼次日的清晨，真正的战场才刚刚开始。前一阶段我们学习了很多理论知识，而离想法落地还有很长一段路要走。政府的理性宣扬奠定了一种基调——一种真正中国化的摩天大楼，既能在体量上反映中华民族的大气，又能包容多元的亚洲文化。我们希望这个奖项与其所选出的优秀高层建筑，能为不远的将来作出指导与贡献。

作为建筑学教授，我常常不满于高层建筑的千篇一律，这是20世纪为保证最佳经济效果所形成的不幸的标准化的产物。这种建筑盛行的时代，甚至造成了城区天际线上布满了天马行空的楼形，只为造成万众瞩目的效果，很少有能与当地社会、环境或历史氛围相契合的。假如一个城市内的建筑照搬到另一个城市，几乎没有人会发现。

从获得成就奖的建筑中我们可以看出，自向世界开放以来到现在成为世界第二大经济体，中国得到了稳步精细的发展，建筑标准从照搬照用到引进专家开展当地建造。例如金茂大厦和台北101大楼这样的主场建筑，它结合了世界专家和当地条件，联结了中国几千年来的建筑史，建造出了这样梦幻般的工程。

最佳高层建筑奖和城市人居奖展示出的近期项目，标志着我们正进入即将到来的"理性繁荣"时代，包括那些看起来像是奇幻随性的建筑，实际上根植于中国深厚的科学环境。我们推举出那些能融合并强化当地环境，而不是互相分离的建筑。在每年有数以百万计的人口城市化的中国，我们希望突出一些示范化的方式，既不是千篇一律的解决方案，也不会华而不实。

根据中国在世界上的影响力，其政府的决策很可能决定着世界未来几年的发展方向。中国担负着很大的责任，也面临着一个巨大的机会。尽管高层建筑只是这场声势浩大的背景下的一个层面，但它是最显而易见的。我们希望继续突出并鼓推那些最有远见的项目和技术，毕竟中国和世界的城市发展正值紧要关头。

The debut of the China Tall Building Awards comes at an auspicious time. For the past 15 years, China has stunned the world with its incredible growth and ambition. The most tangible evidence of this has been on the skylines of its burgeoning cities, which have grown "up" seemingly overnight. The volume of investment and thirst for symbolic representation, or the affirmation of a city's or a leader's prominence in built works, has led many to conclude that China is "an architect's paradise", a boom-town nation, a "free-for-all." Even a few years ago, there would have been much to evidence this claim.

But today, although China still leads the world by a wide margin in terms of skyscraper completions, the atmosphere is somewhat different. Questions abound about the sustainability of the economic boom that has driven this incredible period and the appropriateness of some of the designs that have come to pass.

We have learned much in this heady period, but we still have some distance to go before many of the nascent ideas are truly realized. The call for a semblance of rationality sets a tone that may result in what has eluded China so far – a true Chinese skyscraper vernacular, and within this, many local vernaculars that reflect this nation of a size and diversity that matches entire continents. We hope that this awards program, with its selections of tall-building excellence, provides some additional guidance for the coming years.

As a professor of architecture, I have often railed against the homogenizing effect of skyscrapers, an unfortunate outgrowth of the standardizations and optimizations of the 20th century that made them economically viable. During the boom years, this resulted in urban skylines the world over filled with fanciful shapes that had little to say beyond, "look at me." Most had no relationship to the social, environmental, or historical context in which they were built – pluck one of these commercial monoliths from one city and drop it in another the next day, and few would notice.

With the selections we have made for the Legacy Awards, it is clear there is an arc of rapidly developing sophistication, from the moment of China's economic opening to the world, to its current position as the world's second-largest economy. The architecture has progressed from locally rendered versions of international standards, to "imported expertise and local execution," to projects like the Jin Mao Tower and Taipei 101, masterful projects that bring to bear international expertise and local knowledge to create fantastic buildings that also relate to thousands of years of Chinese architectural history, albeit in a literal way.

The recently completed projects we showcase in the Best Tall Building and Urban Habitat Awards speak to the coming era of what we might call "rational exuberance." Even those that might seem to be arbitrary or fantastical shapes are in fact rooted in deep environmental science and enduring Chinese proverbs. We are holding forth buildings that incorporate and enhance their local environments, rather than stand apart from them. In a country where millions of people are urbanizing every year, we hope to highlight some exemplary ways of going about it, neither cookie-cutter solutions nor gaudy jewels that could be dropped anywhere in the world with the same effect.

Simply as a function of its scale, the decisions China will make in the next few years will largely shape the world. It's an enormous responsibility and a grand opportunity. Although tall buildings are only one dimension of this massive phenomenon, they are among the most visible. We hope to continue to highlight and extoll those projects and technologies that are the most visionary, as the very future of urban life in China, and around the world, is at stake.

# 前言 | Introduction

## 首届中国高层建筑奖联合评审委员会
## The Inaugural China Tall Building Awards Jury

就在中国连续8年成为世界上竣工高层建筑（200 m以上）最多的国家之际，中国最佳高层建筑奖颁奖典礼首次举行。2015年，200 m以上的高层建筑有58%已经在中国竣工，这个数字目前已经超过了大部分国家。正如中国的其他事物一样，其高层建筑也如雨后春笋般涌出。中国的城市移民，是世界史上最大规模的人类迁徙。但显然，我们并不能假设，高层建筑的建造速度能跟得上这个节奏。有迹象表明，中国城市化带来的财富激增和经济迅猛发展的速度已经开始趋缓。

但是，如果中国的高层建筑现象值得研究的话，则不仅仅是因为其速度与规模。中国最佳高层建筑奖项的诞生，不单是为了囊括中国出现的高层建筑的种类和体量，也是为了挑选出高质量的项目，为了强调最佳高层建筑不只是"典范"，也与它们所在城市融为一体。同样需要强调的是，在这些令人眼花缭乱的建筑结构背后，蕴藏着设计师们的匠心独运，这些结构设计本是为了解决某些特定的问题，但对国际社会而言，也有较高的研究价值。

在超过80个参赛建筑中，我们发现一些精选出来的优秀项目。从这些项目中，不难看出，建筑界有些流行趋势比那些只能展示建筑物规模与体量的报告更具有教育意义。

**文脉主义**

贯穿所有参赛作品的主线之一，就是一种对中国自20世纪70年代末期实行改革开放政策以来，不断上升的国际地位和日益增强的国际影响力的认可。我们可以看到，最佳高层建筑奖得主之一，上海外滩SOHO是经典的高层建筑，它同时涵盖了上海外滩的历史与周围的条形弄堂，且它与纽约洛克菲勒中心相似，也有一种广纳公共空间的姿态。

但上海静安嘉里中心则截然不同，该建筑荣获中国最佳高层建筑城市人居奖，此建筑将多种功能集于一身：世界级办公室，酒店，商业零售和住宅，并巧妙地将它们与历史风貌保护和公众所需公共空间融为一体。

郑州绿地中心·千玺广场又是另一番景象。它采用了古代宝塔式的结构，遵循历史悠久的日光集中原则，使用了高度机械化的表面与日光反射装置。当人们明白了其中原理之后，就会知道它那非常规的正十六边形结构正体现了设计师的匠心。

尽管今年很多参赛作品的设计如同寓言一般，但对它们的选择和评定绝非草率。例如位于北京的人民日报总部就很好地体现了其使用者的哲学理念，重新定义了陶瓦的用途，并展示了该项目对楼层空间的充分利用。

The launch of the inaugural China Tall Building Awards program coincides with China's eighth year in a row as the leading country in terms of completions of tall buildings of 200 meters or higher. In 2015, 58% of all tall buildings of this height were completed in China, by far outstripping any other nation, and indeed, most of the continents, in terms of output. As with just about everything else, China's tall building boom is cast in superlative terms. China's urban migration has been the largest human mammalian migration in world history. But it is certainly not a given that the pace of construction of tall buildings will be kept in coming years.

But the pace and scale of construction are not, and truly never have been, the only characteristics that make the Chinese tall building phenomenon worthy of exploration. This program was called into being not just to embrace the incredible volume and variety of tall buildings that have emerged here, but to single out projects of exemplary quality, and to reinforce that the best tall buildings are not merely "icons," but integrated parts of the cities they inhabit. It also emphasizes that there is real ingenuity behind the fantastic structures we are seeing, applied to solve particular issues but highly exportable to the global community.

Across more than 80 entries, in terms of the selected and highlighted projects, we can identify some trends that may be at least as, if not more instructive than reportage of sheer scale and volume.

## Contextualism
One of the strongest threads running through the entries is a kind of localism that nevertheless acknowledges China's ascendant place on the global stage, as much as it does the international influences that have streamed into the nation since the economic opening of the late 1970s.

We can see it in the way that Bund SOHO, a Best Tall Building Excellence Award winner, refers simultaneously to the colonial history of the Shanghai Bund, the historic bar-shaped townhouses of its surroundings, and public-space-embracing gestures of classic tall building complexes like Rockefeller Center. It's also evident in a quite different way at Shanghai's Jing An Kerry Center, the China Urban Habitat Award winner, where the demanding paradigms of world-class office, hotel and retail accommodation have pleasantly intertwined with imperatives of historical preservation and public space that families find usable.

In an entirely different way, the Zhengzhou Greenland Plaza uses an ancient form of the pagoda, and time-honored principles of daylight penetration in combination with a highly engineered façade and a heliostat. When this is understood, the non-conventional, hexa-decagonal shape is revealed as a well-considered, if not obvious design choice.

Even when the designs are allegorical, this year's selections are anything but arbitrary. The People's Daily headquarters in Beijing embodies the philosophy of its occupants, repurposes an ancient terra-cotta cladding practice in a new and exciting way, all while delivering a project of unusually high floor-area efficiency.

## Construction Innovation Enabling Architectural Quality
One of the unique facets of China Tall Building Awards program is the Construction Award, which was launched in order to recognize the incredible achievements in construction that are happening in China's tall buildings – and here again, not just in terms of volume or scale.

The winning project, Forum 66, employs several significant techniques in order to accomplish the feat of raising an architecturally significant supertall building in the harsh climate of Northeast China. Its emphasis on worker safety and safeguarding against compromised structural performance defies the notion that this growth economy favors a construction culture of "speed above all else." Across the Construction Awards winners and nominees, dozens of patents have been awarded, providing some outside validation to the strong impression made upon the jurors.

In the Innovation category, some entries are technologically sophisticated, while others simply reflect above-average, enlightened design thought. The Innovation winner, the curtain wall system at Shanghai Tower, redefines and expands the meaning of "curtain wall," and has given the world one of its most distinctive skyscrapers, in addition to bringing

**建设创新提升建筑质量**

中国最佳高层建筑奖特别的一面在于设立了"建造奖",该奖项的设立是为了对中国高层建筑的杰出成就给予认可,其标准并不是简单的建筑体积或规模。

获奖项目之一,沈阳市府恒隆广场的设计者为了在东北恶劣的气候条件下建起这栋超高层建筑,采用了多项重要技术。该项目强调工人安全第一,加强结构性能优先,反对"速度优先"的建设文化。此次奖项的获奖者与提名者中,多项专利都受到了表彰,给评委们留下了深刻的印象。

在创新奖中,有些作品展现了复杂深奥的技术,有些作品则展示了设计师的奇思妙想。创新奖得主,上海中心大厦的幕墙系统重新定义了什么是幕墙。该项目为世界展现了一栋独特的摩天大楼,以及"垂直城市,空中花园"

**Previous Spread | 前页**

Left: 2016 China Tall Building Urban Habitat Award Winner, Jing An Kerry Center
左图: 2016年中国高层建筑城市人居奖得主,静安嘉里中心

**Current Spread | 本页**

Top: 2016 China Tall Building Construction Award Winner, Forum 66
上图: 2016年中国高层建筑建造奖得主,沈阳市府恒隆广场

Bottom: 2016 China Tall Building Innovation Award Winner, Mega-Suspended Curtain Wall, implemented at Shanghai Tower, Shanghai, 2015 (632 m / 2,073 ft)
下图: 2016年中国高层建筑创新奖得主,上海中心大厦巨型悬挂幕墙,于2015年在上海中心建成并投入使用

的理念。另一个坐落于地震高发区的建筑项目，重庆来福士广场的雄心壮志则体现在其混合结构系统之中，一座弧形水晶廊桥连接着三座塔楼，而塔楼本身也呈弧形。

有些作品虽然采用了较为常规的手法，但也能创造出庄严雄伟的效果，比如深圳太平金融大厦（最佳高层建筑奖得主之一），该建筑采用的是在中国最为普遍的结构系统——双筒结构，但设计师对这种结构的运用上升到了艺术层面，它将自然采光变成了建筑系统的一部分。

### 持续改进

因为这是首次举办中国最佳高层建筑奖，奖项组织方认为，不应漏掉那些自 1978 年改革开放以来改变了中国面貌的地标建筑。中国高层建筑成就奖为我们展示了中国转型成为真正意义上的世界中心过程中，那些塑造中国摩天大楼的巧妙设计。

我们可以清楚地看到，中国是如何从谨慎地参考国际标准，上升到能建造出具有国际竞争力，同时又带有中国特色的伟大建筑。

值得注意的是，一些历史悠久的项目，比如，广州白天鹅宾馆和上海华东电力大楼，在建设之时几乎完全没有西方专家的指导帮助，却十分经久耐用，这些建筑完全可以获得新的租约。较晚竣工的一些获奖建筑，如上海金茂大厦和香港国际金融中心，则一直享受着同行的钦佩和所在城市最高昂的租金。

正如其作品所示，没有人能比此次杰出成就奖获得者汪大绥先生更能代表中国建筑的辉煌成就了。直到今天，他精湛的专业技术和指导仍广受赞誉。正是因为有他，中国才建成了那么多国内乃至世界范围内的标志性建筑。

我们希望在表彰这些项目和个人的同时，也能创造一些自己的价值。我们将筛选出更多优秀的中国高层建筑，让更多的人认识中国高层建筑的杰出成就。我们衷心地欢迎所有阅读此书的人将来都参与到评奖中来，让这个奖项更加全面地记录建筑发展史上的壮丽时刻。

（执笔人：丹尼尔·萨法里克，CTBUH 中国办公室总监）

the much-fantasized idea of "vertical cities with sky gardens" that much closer to reality. The Hybrid Outrigger system, an Honorable Distinction recipient, grounds in reality the fantastic ambitions of the Chongqing Raffles City project, which has an incredible curving skybridge linking its three towers, which also bend out of plane – in a high seismic zone.

In some cases, a highly conventional or popular approach has been manipulated for a sublime result, such as at the Taiping Finance Tower in Shenzhen, a Best Tall Building Honorable Distinction recipient. Here, the ubiquitous double-tube structural system used throughout China has been elevated to an art form that streams natural light into the depths of the disciplined grid of the building.

### A Continuum of Improvement

As this was to be the first China Tall Building Awards, the organizers felt it would be remiss not to recognize some of the landmark projects that transformed China's skylines since the economic opening of 1978. The Legacy Awards thus showcase some of the incredible variety and design ingenuity that characterized the country's skyscrapers as it (rapidly) transitioned from a politically and economically isolated "island" into the de-facto "center of the world." The rites of passage from mimicry of international standards towards a globally competitive quality of buildings with "Chinese characteristics" are clearly observable here.

It's especially noteworthy that some of the oldest projects, such as the White Swan Hotel in Guangzhou and the Huadong Electrical Power Distribution Building in Shanghai, were constructed in the near-total absence of Western expertise, and have proven durable enough to be deemed worthy of top-to-bottom renovations and new leases on life. Later honorees, such as the Jin Mao Tower in Shanghai and the International Finance Center in Hong Kong, continue to command the admiration of their peers and some of the highest office rents in their respective cities.

As his entry will clearly demonstrate, perhaps no one represents this continuum better than Dasui Wang, recipient of the Outstanding Achievement Award. His expertise and guidance, still sought after today, have been instrumental in realizing some of the most iconic structures in China, if not the world.

In presenting these projects and people, we hope that we will be establishing something of a legacy of our own, by becoming the voice and the filter through which the incredible achievements of China's tall building industry are communicated and clarified. In future years, we will welcome all who read this to participate in the program, so that it can become an even more comprehensive record of this amazing time in the history of the built environment.

(Written by: Daniel Safarik, *Director, China Office, CTBUH*)

# China Best Tall Building
## 中国最佳高层建筑奖

# Awards Criteria
## 评选标准

This award recognizes projects that have made extraordinary contributions to the advancement of tall buildings and the urban environment, and that achieve sustainability at the highest and broadest level. The projects must also exhibit processes and/or innovations that have added to the profession of design and enhance the cities and the lives of their inhabitants.

1. The project must be physically located in the Greater China region, including Hong Kong, Macau, Taiwan and Mainland China.

2. The project must be completed (topped out architecturally, fully clad, and at least partially occupied) no earlier than January 1st of the previous year, and no later than the current year's submission deadline. (e.g., for the 2016 awards, a project must have a completion date between January 1, 2014 and December 14, 2015.)

3. The project must be considered a "tall" building. If a project is less than 100 meters, it is unlikely to qualify.

4. The project advances seamless integration of architectural form, structure, building systems, sustainable design strategies, and life safety for its occupants. The project achieves a high standard of excellence and quality in its realization. The site planning and response to its immediate context ensure rich and meaningful urban environments. The contributions of the project should be generally consistent with the values and mission of the CITAB and CTBUH. The project should exhibit sustainable qualities at a broad level:

- Environment – Minimizes effects on the natural environment through proper site utilization, innovative uses of materials, energy reduction, use of alternative energy sources, and reduced emissions and water consumption.
- People – Has a positive effect on the inhabitants and the quality of human life.
- Community – Demonstrates relevance to the contemporary and future needs of the community in which it is located.
- Economic – Adds economic vitality to its occupants, owner, and community.

5. The jury selects a number of "Excellence" recipients, which go on to compete for the China Best Tall Building – Overall Winner Award. Representatives of the Excellence recipients present their projects at the annual Awards Symposium event, after which the jury convenes to determine the overall winner, which is announced at the Awards Ceremony. In addition, the jury selects a number of Honorable Distinction recipients.

本奖项旨在奖励那些对高层建筑的发展和城市人居环境作出较大贡献，以及在实现可持续发展方面具有突出表现的高层建筑项目。申报项目应具有突出的创新能力，并提升城市及人居品质。

1. 该项目的地点应在大中华地区，包括中国大陆和香港、澳门、台湾地区。
2. 该项目必须为已竣工完成的项目（建筑物已经落成，外层装饰全部完成，并至少处于部分使用的状态）。完成时间不早于上一年度的 1 月 1 日，也不晚于今年的申请截止日期（例如，参选 2016 年度的项目，完成时间需在 2014 年 1 月 1 日和 2015 年 12 月 14 日之间）。
3. 参选项目必须是"高层"建筑。如果项目的建筑高度低于 100 m，将不具备参选资格。
4. 参选项目需全面协调和兼顾建筑形式、结构、建筑设备、可持续发展设计和人身安全等方面，该项目的完成过程应体现完美的质量和卓越的标准。项目的总体设计和对周围环境的响应需确保实现积极而有意义的城市人居环境。参选项目应符合 CITAB 和 CTBUH 的价值观与使命理念。可持续发展设计应包括以下方面。
   - 环境：对现场地域合理使用，将项目对自然环境的影响降到最低；使用材料上有创新、节约能源、使用替代性能源、降低排放量和用水量。
   - 人：对居民和人居质量有正面积极的影响。
   - 社区：体现其所在社区的现有需求和未来需求的相关性。
   - 经济：建筑需提升居住者、业主和社区的经济活力。
5. 评委会选出一定数量的优秀奖作品之后，这些获奖作品将继续角逐"中国最佳高层建筑"。获得优秀奖作品团队的代表将在年度颁奖研讨会活动上对项目发表演讲，评委会将在颁奖典礼上评选并宣布获得"中国最佳高层建筑"的作品。此外，评委会还会选出获得荣誉奖的作品。

# China Best Tall Building – Excellence Award
# 中国高层建筑优秀奖

## Asia Pacific Tower & Jinling Hotel
## 金陵饭店亚太商务楼

Nanjing | 南京

The existing Jinling Hotel became a landmark in the center of Nanjing when it was completed in 1982. At the time, it was the tallest building in Mainland China, and quickly became a source of pride for the people of Nanjing. Even today, it is only the seventh-tallest building in Nanjing, but its centrality in the minds of residents and visitors persists.

自1982年建成后，金陵酒店就成为南京市的中心地标。当时，它是中国大陆最高的建筑，建成后很快成为南京人民引以为豪的资本。即使到了今天，虽然它已成为南京第七高的建筑，但无论在当地居民还是外地游客心中，它的中心地位都是不可动摇的。

**Completion Date:** May 2014
**Height:** 242 m (794 ft)
**Stories:** 57
**Area:** 120,000 sq m (1,291,669 sq ft)
**Use:** Hotel / Office
**Owner:** Jinling Hotel Corporation Ltd.
**Developer:** New Jinling Hotel Limited Company
**Architect:** P & T Group; Jiangsu Provincial Architectural D&R Institute Ltd.
**Structural Engineer:** P & T Group
**MEP Engineer:** Jiangsu Provincial Architectural D&R Institute Ltd.
**Main Contractor:** China State Construction Engineering Corporation
**Other Consultants:** Campbell Shillinglaw Lau Ltd. (acoustics); Chhada Siembieda Leung Ltd. (interiors); Meinhardt (façade); Nanjing Institute of Landscape Architecture Design & Planning Ltd. (landscape); Shanghai Citelum Lighting Design Co., Ltd. (lighting); Watermark Associates (way finding)

竣工时间：2014年5月
高度：242 m (794 ft)
层数：57
面积：120 000 m² (1 291 669 ft²)
主要功能：酒店 / 办公
业主：金陵饭店有限责任公司
开发商：新金陵饭店有限责任公司
建筑设计：巴马丹拿建筑设计咨询有限公司；江苏省建筑设计研究院有限公司
结构设计：巴马丹拿建筑设计咨询有限公司
机电设计：江苏省建筑设计研究院有限公司
总承包商：中国建筑工程总公司
其他顾问方：金宝声学顾问公司（声学设计）；Chhada Siembieda Leung 有限责任公司（室内设计）；迈进集团（建筑立面）；南京园林规划设计院有限公司（景观设计）；上海城市之光灯光设计有限公司（灯光设计）；Watermark 设计事务所（标识系统）

Recognizing the need for new, contemporary and extended accommodation, the hotel's owners held a limited design competition to establish a land use and development strategy. The winning proposal houses the new extension with all associated facilities in a single, 57-story-tall tower that rises to a height of 240 meters.

A podium structure related and linked to the existing hotel houses new, extensive banqueting facilities, restaurants and a health club with an indoor swimming pool on its roof. The remainder of the mid-city site is allocated to circulation and extensively landscaped gardens.

The square tower is essentially divided into two sections. The 370 new rooms and suites of the hotel's 5-star extension are provided on the top floors of the tower and reached directly by express elevators. An exclusive lounge located above the guest rooms tops the tower. Class-A offices are housed below the hotel, extending over 30 typical floors, featuring spacious and column-free accommodation.

考虑到扩建出全新的现代住所的需要，酒店业主发起了一个设计大赛以创建一个土地使用和发展的战略。胜出的设计方案将会应用到未来大楼的扩建及一切相关设施的建设上，届时大楼的楼层将多达57层，高达240 m。

酒店的裙楼结构与现有建筑相连，内包含新的餐饮设施、饭店和一个顶部带有室内泳池的健身俱乐部。酒店的其余场地被设计作为交通用地和景观花园。

方形塔楼被分成两个部分。顶层部分是酒店，达到五星级标准，有370个房间和套房，配有直达高速电梯。塔顶客房的上面是一个专属休息室。酒店下面为甲级办公室，有30个标准楼层，以宽敞、无柱的办公空间为特色。

新的扩建建筑坐落在现存酒店的中轴上，保持着原始设计的方形轮廓，隐藏的角落也在中轴上，保持着方形，代表阴阳呼应。在金陵饭店亚太商务楼的顶部是一个迷人的屋顶花园，可以纵观城市全景。往下则是巨大的绿色花园、喷泉、瀑布，在城市中心构成了一个16 000 m² 的绿色空间。

*"The Asia Pacific Tower is particularly commendable in finding a design language that relates successfully back to the original hotel, and also makes an additional level of cultural contribution to its Nanjing context."*

"亚太商务楼的可贵之处在于找到了一种既能成功保持原有的酒店风格，又为南京的文化意蕴锦上添花的建筑语言。"

Design Juror　设计评委

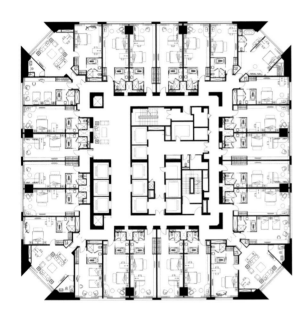

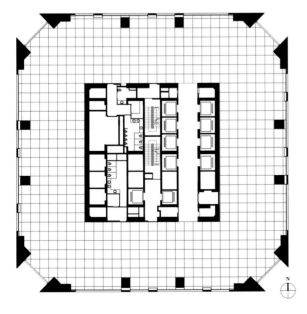

**Previous Spread | 前页**

Left: Overall view from southeast
左图: 东南面全景

Right: Retail arcade on basement level 1
右图: 地下一层商场

**Current Spread | 本页**

Opposite Top: Interior view of banquet room
对侧上图: 宴会厅内景

Left: Typical hotel (top) and office (bottom) floor plans
左图: 酒店标准层 (位于建筑上部) 和办公标准层 (位于建筑下部)

## Jury Statement ｜ 评委会评语

Few projects are as successful as the Asia Pacific Tower & Jinling Hotel at meeting the commercial demands of expansion without sacrificing the original character of the building. The new structures are clearly conceived in the same formal language as the old, highly compatible, yet subtly distinctive. The choice of materials, positioning of new public areas, and the quality of landscaping and exterior detailing all serve to enhance the overall experience. What was Nanjing's tallest building can now be assured of its place in the region's skyscraper vanguard for generations to come.

很少有项目能够像金陵饭店亚太商务楼这样既成功地满足了商业扩张的需求，同时又没有牺牲原有的建筑风格。这一全新的建筑结构很明显在努力地与旧有的建筑语言保持一致，高度兼容，但又创造了微妙的独特之处。材料的选择、新的公共区域的选址以及景观和外部细节处理，都提升了整个建筑的品质。作为南京曾经最高的建筑，如今它已成为几代人心目中这一区域摩天大楼的先锋和典范。

# "Everything about the extension shows a great deal of thought and care went into a project that was meant to carry forth a legacy sustainably."

"任何向项目注入大量思考与关注的提升都意味着能够不断达成某项成就。"

Design Juror    设计评委

**Above:** Interior view of office lobby
上图: 办公大堂内景

**Opposite Top:** Level 2 floor plan
对侧上图: 二层平面图

The new extension is placed on the central axis of the existing development and retains the square shape of the original design, but does so with recessed corners, representing the shape of the Yin in response to the Yang. At the top of Jinling Hotel Asia Pacific Tower lies a beautiful and alluring rooftop garden with panoramic views of the city. Below is a huge landscape of green gardens, fountains and waterfalls, providing a space of some 16,000 square meters in the center of the city.

Basement levels extending over the whole site contain car parking, loading and unloading and back-of-house facilities related to the hotel. Basement Level 1 features an extensive retail arcade lined with shops and restaurants, and is linked to an adjoining mass transit station, providing easy and convenient access for visitors and workers alike.

The redesign and extension of the complex addresses requirements for contemporary comfort and convenience, as well as for energy efficiency, output and productivity standards. Importantly, the end effect of the addition on the original is also an essay on the reconciliation of formal gestures, which channels opposing forms into a soothing harmony. This is reflected in the way the site plan pushes building volumes out toward the compass points at the site's corners, drawing visitors inwards.

The tower is clad with white aluminum panels in combination with double-glazed, tinted windows,

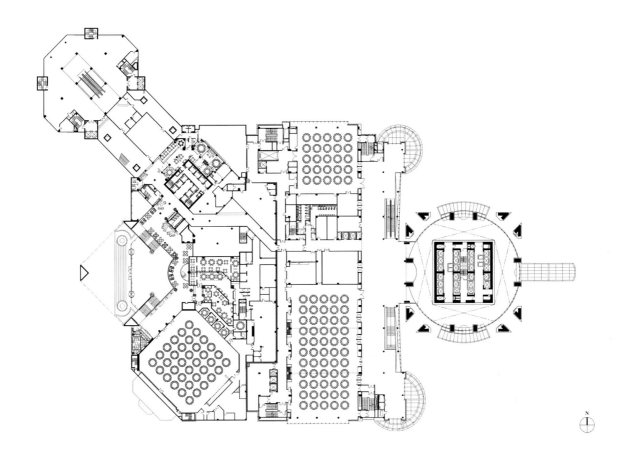

and the podium is lined with beige Brazilian granite throughout. The rigid geometry of the façade is counteracted by the swirling pattern of paving stones around the fountain at the driveway, and in the smooth glass drum inset into the notch at the front entrance, itself a fragment of the overall circle-within-a-square motif of the focal tower. At night, the rounded corners shed strips of light that are suggestive of an energy the square plan can barely contain. Materials are chosen for their complementary nature and rich hues, from wall tapestries through to light fixtures and marble flooring. Artifacts and reveals that showcase the original design's discipline and dignity are displayed throughout both the new and renovated areas.

地下层的范围占满整个基地，内含停车场、货物装卸地、酒店的后勤设施。地下一层是一个零售商场，驻有商店和餐厅，毗邻公交枢纽，为游客和工作人员提供了极为便利的通道。

建筑的重新设计和扩建要求符合现代、舒适和便利的标准，也需要达到能效、产放和生产的要求。重要的是，扩建部分的最终效果也需要与既有建筑协调起来，使它们成为一个和谐的整体。这一点在总平面设计图上有所反映，根据设计图，主体建筑的出入口面向各个方位，以吸引游客进入建筑内部。

塔楼外层以白色铝板、双层彩色玻璃覆盖。裙楼全部以巴西米色花岗岩饰面。车道上喷泉周围漩涡形的铺装中和了建筑立面上严谨的几何感。位于正门处嵌入的玻璃鼓的形态正是原有建筑"方套圈"母题的一部分。夜晚，从圆形角落溢出的条形光喻示着这个方形项目蕴藏的能量。从墙上挂毯、灯具到大理石地面，所有的材料都经过了精挑细选，呈现出相得益彰的自然而饱满的色调。

# China Best Tall Building – Excellence Award
中国高层建筑优秀奖

## Bund SOHO | 外滩SOHO
Shanghai | 上海

The Bund SOHO is located at the southern end of the famous riverfront boulevard of Shanghai - the Bund - and adjacent to a preserved historic block. Marking the transition from the old to the new Bund, SOHO Bund is a mixed-use urban development, comprising offices, retail and entertainment functions. Four high-rise office buildings, with heights between 60 and 135 meters, are presented as elongated, shifted volumes, creating a vivid urban space with small squares and alleys leading to the Huangpu River waterfront of the Bund. The basic layout

外滩SOHO地处上海外滩南端，毗邻历史街区。它标志着老外滩向新外滩的华丽转身，是一个集办公、商业和娱乐功能于一体的多元化发展街区。该项目由4栋60~135m的高层办公楼交错排列组成，小型广场和胡同小巷一直延伸至外滩，创造出生动的城市空间。项目的布局旨在为新外滩建造一座适应所有年龄层的时尚繁华多元化的商业建筑。外滩SOHO致力成为能够给予当地居民归属感的场所，也致力成为各地都市游客经常光顾的地方。

**Completion Date:** 2015
**Height:** 136 m (445 ft)
**Stories:** 31
**Use:** Office
**Owner/Developer:** SOHO China Co. Ltd.
**Architect:** von Gerkan, Marg and Partners Architects; ECADI
**Structural Engineer:** ECADI
**MEP Engineer:** ECADI
**Main Contractor:** Shanghai Construction No.1 (Group) Co., Ltd.
**Other Consultant:** AIM Architecture (interior design of office and commercial area underground)

竣工时间：2015年
高度：136 m (445 ft)
层数：31
主要功能：办公
业主／开发商：SOHO中国有限公司
建筑设计：GMP建筑师事务所；华东建筑设计研究总院
结构设计：华东建筑设计研究总院
机电设计：华东建筑设计研究总院
总承包商：上海建工一建集团有限公司
其他顾问方：恺慕建筑（办公与地下商业空间室内设计）

of this project aims to create a lively, diverse and pleasant destination for people of all ages in the New Bund, a regeneration area south of the city center that has seen rising and falling fortunes in the past. Bund SOHO aims to play a future role in making it a place where people from the local community can have a sense of belonging, and a site where people from other regions or countries want to go.

The design is guided by a deep study of the historic surroundings, such as the typical bar-shaped "Nong Tang" townhouses, the narrow street grids leading to the river, and the skyline of the classic, vertical stone façades of the Beaux-Arts and Art Deco Bund commercial buildings. The site runs much longer from east to west than from north to south, yielding minimal frontage on the most favored (river) side, thus calling for a response that would maximize the sense of exposure to the riverfront while still yielding high-quality spaces deep within the site. Consisting of bar-shaped volumes, offset in plan, the design accomplishes adequate natural light distribution on each floor plate and provides multiple façades and planes that generate visual interest

该设计始于对外滩充满历史感的环境的深入研究，这些历史要素包括了深具上海特色的条形"弄堂"，它们构成了狭长交错的街道网，一直通往黄浦江。江边建筑群构成了上海的特色天际线，外滩商业建筑群的石材垂直立面兼具古典与装饰艺术的美感。项目基地东西向较长，南北向很短，这导致了在该黄金地段上建筑的面江立面非常狭窄，进而要求项目中的建筑需要尽可能地朝向江边，同时也要让基地内侧仍然具有高品质的空间感。建筑呈条形体块，平行布局，设计力图让每一个楼层都能得到充分的自然光照，同时让建筑的立面和平面在白天多个位置上都能够产生丰富多彩的视觉趣味。为了避免让目光只聚焦在一个点上，空间虚实的丰富交互令建筑尤具魅力，使人忍不住要去一探究竟，从建筑的外部、里面或穿越其中，以多个而非单一的视角去感知其中的"尊贵之处"。

本质上，这是一个多渗透和多渠道的设计。为了实现此目标，该项目内还设计了多条步行线路、多个车辆入口和车辆通道。沿着基地边界有三个直通地下停车场的机动车辆出入口和一个非机动车出入口。另外，商业购物街的出入口面向人民路、新开河路和外滩周围街区。以上设计，

## Jury Statement | 评委会评语

Bund SOHO resolves a difficult site, historic surroundings, the requirements of the modern office building, and the responsibility of a high-profile waterfront location in Shanghai with a reserved distinction. It is like a ship's prow, pushing the frontier forward of the established central Bund into new, uncharted territories, with visible confidence. There are discernible shades of Raymond Hood's work at Rockefeller Center, New York and nods to the Art-Deco predecessors on the Bund, as well as to the existing grain of the neighborhood. Nevertheless, a project of unique qualities has been rendered here.

外滩 SOHO 解决了一个老大难的地块，这里处于历史环境中，又有对现代办公建筑的需求，还担负着保护上海这一片极为著名的滨水地区的责任。它就像一艘船的船头，满怀自信地将已有的外滩中心区域推向一个新的、未知的领域。外滩 SOHO 明显受到 Raymond Hood 的纽约洛克菲勒中心的影响，但它也从周围的环境中汲取营养，向外滩的 Art-Deco 先驱们致意。尽管如此，这个项目还是在这里向我们呈现出其自身独特的风采。

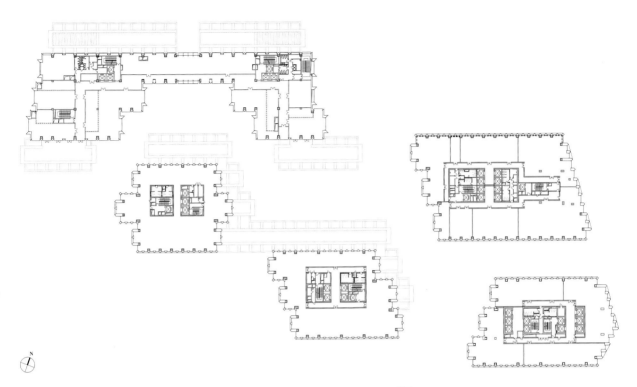

**Previous Spread | 前页**

Left: Overall view
左图：全景

Right: Interior view of office lobby
右图：办公大堂内景

**Current Spread | 本页**

Opposite Bottom: View from retail-lined courtyard
对侧下图：商场庭院

Top: Typical floor plan
上图：标准层平面图

Left: Pathway leading to the Bund waterfront
左图：通向外滩的步道

*"It's clear there are classic allusions in the form and organization of the towers, but it's equally clear that the time was taken to develop a site-specific solution that draws the eye from multiple angles."*

*"该建筑的形式和组织方式明显遵循了经典的范式，但同时也可以清楚地看到，从各个角度上，这个历时良久、因地制宜的设计方案都光彩夺目。"*

Design Juror　设计评委

at many points during the day. Instead of focusing all eyes on one point, the interplay of void and solid entices exploration, such that many views out of, into and through the complex feel "privileged."

In essence, this is a multi-penetration-and-channel plan. In order to achieve the above target, multiple possibilities for pedestrian and vehicular entry or passage through the site are created. Around the site, there are three motorized-vehicle entrances/exits directly leading to the basement and ground level drop-off zones, and one non-motorized-vehicle entrance/exit to the basement. In addition, the entrances/exits of the commercial shopping street are located along the perimeter blocks of Renmin Road, Xinkaihe Road and the Bund. The overall traffic flow of the surrounding area, which makes access both convenient and fast, enhances the likelihood of success for the commercial and entertainment functions, and creates an abundant, diverse, forward-looking and dynamic urban environment focused on people.

> "Unlike many large-scale urban projects, which are often neglectful of their urban surroundings, Bund SOHO takes a much stronger cue from the inviting nooks and crannies of the local vernacular lanes."
>
> "许多大型项目常常忽略周边的城市环境，但外滩 SOHO 却从当地的大街小巷各个角落中攫取到了灵感。"
>
> Design Juror　设计评委

**Opposite Top:** View of the pedestrian retail corridor
对侧上图: 商场步行街街景
**Opposite Bottom:** View of the pedestrian retail corridor and sub-level commercial space
对侧下图: 地下商场与步行街街景
**Left:** Overall view of the complex from across the Huangpu River
左图: 外滩与外滩SOHO

The main spatial concept in the plan is to arrange the towers A, B, C, D and E along the east and south sides of the site, with Building F occupying the northwest side and Building G on the north side. In the west, an outdoor commercial shopping street with three floors extends from the corners defined by Yong'an Road and Renmin Road to the Bund, covering the whole length of the site. Many public gathering spaces of different sizes and forms are arranged along this shopping street. Each space having its own scale and features, these spaces provide stages for various types of public activity throughout the year.

Overall, the effect is certainly a case of becoming something more than the sum of its parts. In addition to its local Shanghai references, the serrated, radiator-grille-like forms, setbacks, and bold geometries of the project, framing interlaced public spaces, recall something of New York's Rockefeller Center, albeit scaled down to reflect its local surroundings.

使周边地区的车流变得方便快捷，强化了外滩 SOHO 的商业和娱乐功能，创造出了一个丰富多样、以人为本、具有前瞻性的动态城市环境。

设计的主要空间概念是：把 AB、C、D、E 四栋主楼分别安排在基地的东侧和南侧，F 楼在基地西北处，G 楼则在基地正北处。基地西侧，建有一条三层楼的室外商业购物街，它从永安路、人民路交叉口一直延伸到外滩，涵盖整个基地的宽度。沿街分布着各种不同大小、不同形态的公共休闲区。每个区域规模各异，各具特色，为开展各类公共活动提供了良好的平台。

总体来说，外滩 SOHO 的整体效果超越了各个部分之总和。除去上海的特有风格，其锯齿状的、如散热器格栅样的形态，建筑后退线，大胆的几何构成，这些元素把各个公共空间完美交织在一起，大有纽约洛克菲特中心的格调，而相对较小的建筑体量却又折射出外滩当地的环境特色。

# China Best Tall Building – Excellence Award
# 中国高层建筑优秀奖

## Hongkou SOHO | 虹口SOHO
Shanghai | 上海

Located at the crossing of Wusong and Wujin roads, Shanghai, Hongkou SOHO is well-positioned from a commercial standpoint, but it is on a challenging site that demanded a creative response. Located in the burgeoning North Sichuan Road area near the intersection of Metro lines 4 and 10, with several heritage buildings and a park nearby, the building and its grounds are in some ways being pulled in several directions at once. This multi-directionality has proven a central and valuable determinant of the design.

从商业角度看，位于上海市区虹口区吴淞路和武进路十字路口的虹口SOHO选址极好，但这也是一块亟需创意，充满挑战的地块。项目位于高速发展的四川北路，临近地铁4号线和10号线的交汇处，周边有多个保护建筑，附近还有一个公园。这些条件在不同空间方向上影响着该项目的场地与建筑。这种多方向性的特点最终被证明成为设计的核心价值。

**Completion Date:** August 2015
**Height:** 134 m (438 ft)
**Stories:** 29
**Area:** 60,943 sq m (655,985 sq ft)
**Use:** Office
**Owner:** SOHO China Co. Ltd.
**Developer:** Shanghai Xusheng Property Co., Ltd.
**Architect:** Kengo Kuma and Associates; Tongji Architectural Design (Group) Co., Ltd.
**Structural Engineer:** Tongji Architectural Design (Group) Co., Ltd.
**MEP Engineer:** Tongji Architectural Design (Group) Co., Ltd.
**Main Contractor:** Shanghai Construction No.1 (Group) Co., Ltd.

竣工时间：2015年8月
高度：134 m (682 ft)
层数：29
面积：60 943 m² (655 985 ft²)
主要功能：办公
业主：SOHO中国有限公司
开发商：上海旭升置业有限公司
建筑设计：隈研吾建筑都市设计事务所；同济大学建筑设计研究院（集团）有限公司
结构设计：同济大学建筑设计研究院（集团）有限公司
机电设计：同济大学建筑设计研究院（集团）有限公司
总承包商：上海建工一建集团有限公司

The tower is covered with unique vertical shading and 18-mm-thick white aluminum strips knitted as a lace net, providing various façade expressions according to sunlight and shade. The difficulty of including parking and vehicle access on the asymmetrical site has led the designers to set the building back from the street, making it easier to admire from a distance, while softening its presence on the street edge, an effect that gains much assistance from the extensive planting around the immediate perimeter.

In contrast to the hard and cold image of conventional tall buildings, the curved shading system, shaped as a triangle joint in plan and gradually changing shape as it gains altitude, forms a constructed element with rhythm and a façade with topographical characteristics The building appears from some solar and viewing angles as a solid and powerful being, and from others as a soft and delicate thing. Substantial support columns are reduced in prominence as the line of glass enclosure dodges around them. The metal net turning up to arch over the main entrance resembles a lady's dress flowing in the gentle breeze. The paving stones of the interior atrium are curved to correspond with the building façade, to give the image of soft cloth. The net-like motif is repeated on the interior, at some points hanging from the ceiling like a drop curtain or a low-flying cloud. On higher floors, the resulting sense of enclosure is at some points juxtaposed with the thrill of flying, a sensation enhanced by floating stairs and double-height spaces. In the lobby, the essence is of walking into an ice cave of calm and reserve after leaving the thrum of the street outside.

The project is certified LEED Gold, providing quality office space for the commercial area of North Sichuan Road. The curtain wall uses low-E insulating glass with sealant, which satisfies both energy-saving and building-sfatey requirements. The metal shading system and façade aluminum strips provide the building with adequate protection and lower the sunlight reflection from its surroundings without affecting the view from indoor rooms. The landscaped roof and podium reduce energy loss off

整幢建筑采用独特的垂直感外遮阳百叶，以18 mm宽的白色铝条编织成具有通透感，犹如蕾丝网格般的百叶，给予建筑不断的明暗变化。根据阳光的角度、强度及颜色的变换，建筑立面的纹理也随之产生不同的光影变化，时而锐利，时而柔和。这一弯折百叶系统在平面上成三角形拼接，在高度上逐渐变化，形成有韵律的构件，借此创造出与一贯以坚硬冰冷素材所组成的高楼所截然不同的、更为柔性的建筑，如同女士时装般的细腻。室内大厅则用石材作出与立面一致的摺叠纹理效果，让一向具有沉重感的石材也变化出像衣料般轻柔的幻觉，如同被施了魔法一般。

在不规则形基地里，停车场和车辆出入口的位置变化让设计者决定将建筑位置退后、远离道路，由此，建筑获得了更多的观赏视距。同时，建筑周围大量的绿化也柔化了建筑的街道立面。

与又冷又硬的传统建筑对比，这套曲线遮光系统在平面上呈三角形连接，随高度变化而逐渐改变，形成一组有节奏的构件，它们也使幕墙具有地貌般的特色形态。这座建筑某些光线条件和角度看，是坚固而有力的存在，但从

**Previous Spread | 前页**
Left: Overall view
左图：全景
Right: Detailed view of the vertical shading on the west façade
右图：西立面上的建筑细部

**Current Spread | 本页**
Opposite Bottom: Building entrance
对侧下图：入口
Left: Interior view of the podium level common space
左图：裙房公共空间内景
Above: Typical floor plan
上图：标准层平面

*"The design moves of the tower's façade draw attention, but for all the right reasons, including solar protection."*

*"建筑立面的动感设计引人注目，而且还具有遮阳的实质作用。"*

Design Juror 设计评委

**Top:** Interior view of the grand staircase and common area
上图：公共空间内景

**Right:** Overall view of the tower in its historic context
右图：街区文脉

**Opposite Top:** Interior view of the building entrance
对侧上图：入口内景

# "The use of a relatively conventional material in an unconventional way has brought myriad benefits to Hongkou SOHO as well as the surrounding community."

"以非常规的方式对常用材料加以应用，使虹口 SOHO 项目和周边社区受益匪浅。"

Design Juror　设计评委

the top-floor air conditioners and the urban heat island effect, while providing a good view for the office tower. Potable water is conserved through the use of a greywater system for flushing and irrigation, as well as high-efficiency appliances.

## Jury Statement | 评委会评语

Like a starlet of film's golden era or an exuberantly striped tuxedo, the Hongkou SOHO tower manages to absorb and reflect back the glare of commercial imperatives with a winning elegance. Its fundamental enchantment is in what it reveals and does not reveal. It is smooth and dignified, without being cold or severe. The modification of a rectilinear typology to accommodate an awkward site has resolved beautifully. And though its striking façade is what captivates on first view, like the most enduring icons, the overall design has enough integrity and depth to keep offering new surprises to frequent visitors.

它像黄金时代的电影明星，或像夺目的条纹礼服，虹口SOHO大厦成功地以极致优雅的姿态蕴含并反射出商业运行的光芒，其根本魅力就在于它所展现出来的以及它所隐藏的各种美感。它是柔滑而优雅的，完全没有冷酷或严苛的感觉。直线设计的改良方案非常漂亮地解决了因为尴尬的地理位置而带来的难题。尽管建筑的外立面很引人注目，但作为一个长久性的地标建筑，其整体设计具有足够的完整性和深度，能不断地为经常造访这里的人群提供不断的惊喜。

其他的角度看，又变得柔和而精致。玻璃幕墙将外突的结构立柱围合起来。主入口拱形的金属网像少女的裙摆在微风中飘扬。场地景观的石子路弯弯曲曲与幕墙形式呼应，有种织物的感觉。内部也使用网状主题，天花板上的悬挂物像一段垂落的窗帘，或一朵低飞的云。在高层区域，围栏的最终视觉效果让人们有种踩在云端的快感，悬浮楼梯和挑空空间强化了这种感觉。当从外面街道走进大堂时，则会感到安静和舒适。

这个项目获得了 LEED 金奖认证，为四川北路的商业地区提供了优质办公空间。幕墙采用断热铝合金框、中空夹胶 LOW-E 玻璃，这不仅满足了节能需求，也保证了建筑安全性。金属遮阳系统和立面铝条为建筑提供了充分的保护，也减少了该建筑周围建筑的反射光影响，却不影响室内视野。裙房的绿化屋面减少了顶层中央空调的能量散失和城市热岛效应，同时也为办公楼提供了良好的景观。中水系统用于冲洗和灌溉植被。建筑同时还设有节水装置以节约饮用水。

# China Best Tall Building – Excellence Award
# 中国高层建筑优秀奖

## Wangjing SOHO | 望京SOHO
Beijing | 北京

The Wangjing SOHO Project is designed as three dynamic mountain- or fish-like forms, pulling flow through the site with their convex forms. The juxtaposition of the towers affords a continuously changing, elegant and fluid view from all directions. The exterior skin of the towers consists of flowing, shimmering ribbons of aluminum and glass that continuously wrap around the buildings and embrace the sky, threading through a landscape with approximately 60,000 square meters of green area

望京 SOHO 项目的形态就如三座连绵起伏的"山峰",宛若从基地中涌动而起。多个大厦的并置带来连续性的变化感,从各个角度来看都有一种优雅的流动美。楼体外表皮由流畅且闪闪发光的铝板和玻璃构成,如同丝带一般内裹大楼而外拥天空,基地内还有面积约 6 万 m² 对外开放的绿色景观。该项目的设计灵感来自周边的都市动感,以及太阳与风的自然气息,旨在成为望京地区的地标,作为一个窗口或灯塔,可以让来往于高速公路或北京首都国际机场的乘客看到。

**Completion Date:** 2014
**Height:** Tower 1: 118 m (387 ft); Tower 2: 127 m (417 ft); Tower 3: 200 m (656 ft)
**Stories:** Tower 1: 25; Tower 2: 26; Tower 3: 45
**Area:** 123,573 sq m (1,330,129 sq ft)
**Use:** Office
**Owner/Developer:** SOHO China Co. Ltd.
**Architect:** Zaha Hadid Architects; CCDI Group
**Structural Engineer:** China Academy of Building Research; CCDI Group
**MEP Engineer:** Arup; CCDI Group
**Main Contractor:** China State Construction Engineering Corporation
**Other Consultants:** Arup (façade); Ecoland (landscape); Environmental Market Solutions, Inc. (LEED); Ikonik (way finding); Inhabit Group (façade); Lightdesign (lighting); Yonsei University (wind); Zaha Hadid Architects (landscape)

竣工时间:2014 年
高度:塔楼 1:118 m(387 ft);塔楼 2:127 m(417 ft);塔楼 3:200 m(656 ft)
层数:塔楼 1:25;塔楼 2:26;塔楼 3:45
面积:123 573 m²(1 330 129 ft²)
主要功能:办公
业主 / 开发商:SOHO 中国有限公司
建筑设计:扎哈·哈迪德建筑事务所;悉地国际
结构设计:中国建筑科学研究院;悉地国际
机电设计:奥雅纳工程咨询有限公司;悉地国际
总承包商:中国建筑工程总公司
其他顾问方:Arup(外立面);易兰(景观);EMSI(LEED);依科(标识);英海特工程咨询集团(立面);Lightdesign(照明);延世大学(风工程);扎哈·哈迪德建筑事务所(景观)

*"Wangjing SOHO somehow manages to reconcile the excitement and energy of a great city with the need to be near to sunlight and greenery."*

"望京 SOHO 借助与阳光及绿色的亲密接触出乎意料地达成了一座大城市的激情与能量的融合。"

Design Juror　设计评委

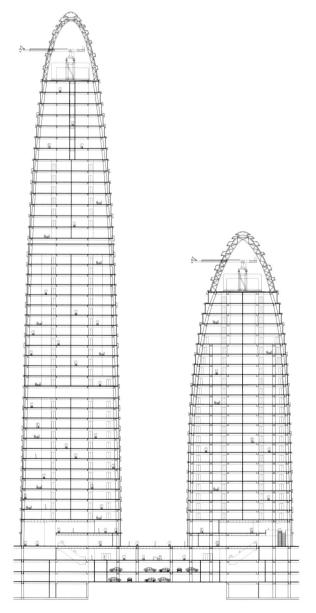

## Jury Statement | 评委会评语

In a city known for its regimented street grid and rectilinear building organization, Wangjing SOHO takes an unusual kink in the grid, forming a semi-circular site, and exploits it to the fullest. The façades ripple and buzz, pathways spin out like traced rays through green surrounds, and from every angle, the fish-eye, fun-house morphology looks different. As such, the shimmering edifice presents a compelling, if contrary, rip in the fabric of the street grid's efficiency, making us question what we value in tall building design, and ask if there might be something more.

在一个以严谨的街道网格和直线形建筑著称的城市，望京 SOHO 以一个不寻常的扭曲形态，塑造了一个半圆形的基地，并将它充分开发出来。建筑外立面看起来宛如波纹一样，通道是旋转的，看起来就像是通过周围的绿植散发出来的一样，而且从不同的角度看，这座有趣的建筑都像是从鱼眼镜头看到的那样具有不同的形态。这座闪闪发光的大厦十分具有吸引力，它突破了严谨的道路网格结构，让我们开始思考高层建筑的设计价值，并扪心自问是否还有可能创造出"更多的东西"。

**Previous Spread | 前页**

**Left:** Overall view
左图：全景

**Right:** Detail view of the aluminum and glass ribbons on the north façade
右图：北立面上铝板与玻璃的"彩带"细部

**Current Spread | 本页**

**Opposite Left:** Interior view of the fluid ceiling design
对侧左图：室内流动的天花板造型

**Opposite Right:** Building sections
对侧右图：建筑剖面图

**Right:** View of public landscape in the north plaza
右图：北广场的公共花园景观

open to the public. Inspired by the surrounding movement of the city, and the sun and wind, the project aims to lend a strong identity to the Wangjing area, creating a gateway / beacon that can be seen by travelers along the highway heading to or from Beijing Capital International Airport.

The site for the Wangjing SOHO Project is located in the Chaoyang District of northeast Beijing, between the Fourth and Fifth Ring Roads. The area contains the offices of many Chinese startup companies, as well as global companies such as Microsoft, Daimler, Caterpillar, Panasonic, Nortel and Siemens. It is conveniently located on the way to the airport and near various metro stations, and is home to a vibrant mix of local and international residents and visitors.

The building program contains offices and retail above grade, retail below grade in B1 basement level, and parking and mechanical equipment in the B2, B3 and B4 basements. The composition of the towers extends into the surrounding landscape, with flowing lines creating paths of movement and exciting activity zones of shopping and leisure. The lines of movement extend to the perimeter and integrate all the green areas around the site. Between the main building towers is a "canyon" of retail and several pavilion gate buildings that create a shopping street at the ground level. There are two sunken garden courts east and west of the canyon that continue the landscaped paths down to the retail concourse below.

望京 SOHO 项目位于北京东北部的朝阳区，在四环和五环之间。该区域设有许多中国创业公司以及跨国公司的办公楼，如微软、戴姆勒、卡特彼勒、松下、北电和西门子。它交通便利，处于通往机场的路上，靠近多个地铁站。这里生机勃勃，还是当地人、外国人和游客的居住区。

项目内包含办公室和地上的零售区，地下的零售区位于地下 B1 层，停车场和设备室位于地下 B2、B3 和 B4 层。建筑携同流线形的步行道路、振奋人心的商业休闲活动空间一起，延展到周边的景观中。动感的线条一直伸到基地周边，与建筑周围的绿地相整合。主楼之间有多个零售商店和活动区，它们与几个入口构筑物共同形成了一条"峡谷"形的地面购物街。在"峡谷"的东部和西部各有一个下沉式花园，景观道路一直延续到下沉购物广场中。

The main tower entrance lobbies, facing outwards to the city, welcome visitors into large dynamic halls that direct one into the office tower floors above, and to the breezeway and retail levels at the second floor and sunken garden levels below. Up above in the office towers, there are simple open-plan office spaces, offering natural daylight and continuous panoramic views in all directions.

Most of the roofs are covered with louvers and the tops of the roof surfaces are coated with highly reflective material, in order to mitigate the heat-island effect in the city. The buildings have double-insulated unitized glazing systems and horizontal bands of white aluminum that provide overhangs for sun shading, doubly functioning as maintenance terraces and water collection devices.

To encourage more sustainable transportation access, special parking spots are reserved for low-emission cars. Bicycle parking and shower facilities are also provided. Direct access to subway stations and bus stops nearby has been integrated into the planning.

For better indoor environmental quality for the occupants, the fresh air exchange rate per person exceeds the ASHRAE standard by 30%. Highly efficient filters are installed to remove 90% of the PM2.5 particles in the air-conditioning system. In the interior design, low-volatile organic compound (VOC) materials are carefully chosen to eliminate pollution from the outset.

**Right:** Interior view of the two-story lobby
右图: 两层高的大堂内景

**Opposite Left:** Overall view from the north
对侧左图: 北部全景

**Opposite Right:** Typical floor plan
对侧右图: 标准层平面图

面朝城市街道的主楼入口大堂，迎接游客进入宽敞的动感大厅，该厅可直接向上通往办公楼、二楼的天桥和零售店，向下通往下沉式花园。在办公楼中，有简单的开放式的办公空间，提供自然采光和连续的全景视野。

为了缓解城市的热岛效应，大部分的屋顶覆盖着百叶，顶层的屋顶表面覆盖着高度反光的材料。建筑遮阳采用单元式的双层隔热低辐射玻璃系统和水平带状的白色铝制外幕墙板，还配有维护平台和雨水收集装置。

为了鼓励更为可持续的交通方式，项目中还专门规划了低排放汽车的停车场。此外还有自行车停车场。该项目还规划可以直接与周边的地铁站和公交车站对接。

项目为入驻者提供了良好的室内环境，所提供的人均新风量超过 ASHRAE 标准 30%。安装在空调系统中的高效过滤器可滤去空气中 90% 的 PM2.5 颗粒。在室内设计中，为了从源头消除污染，精心挑选了挥发性有机化合物 (VOC) 含量低的材料。

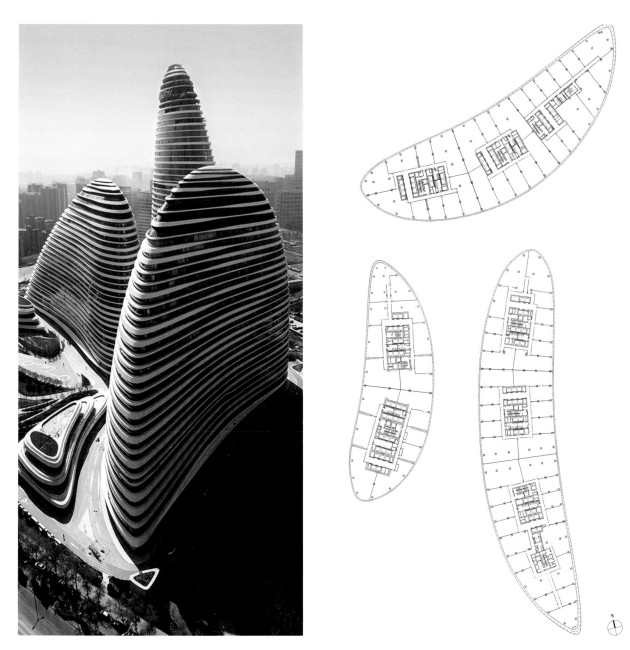

*"The project is both 'organic' and 'otherworldly,' and, unlike so many buildings, seems to contain mysteries and subtleties of detail that will reveal themselves over time."*

"项目既'原始'又'超脱',而且,并不像许多建筑,它似乎有着饱含细节的神秘与微妙,这将随着时间的流逝而彰显本色。"

Design Juror　设计评委

# China Best Tall Building – Honorable Distinction
# 中国高层建筑荣誉奖

## Fake Hills Linear Tower | 假山大厦
Beihai | 北海

The vast majority of development in China's new cities takes the form of residential housing, often standardized and cheap, to guarantee a quick return on investment. The coastal city of Beihai is highly representative of this tendency, which is all the more unfortunate, given its privileged geographical position. Confronted with an 800-meter-long waterfront site, the designers of Fake Hills asked, "Is it possible to build high-density, economically viable housing that is also architecturally innovative?"

Despite its apparently tongue-in-cheek name, Fake Hills presents a genuine and authentic hybrid experience between high-rise life and a seaside park, and between a residential tower and residential compound typology. Combining the best of both typologies, providing for oceanfront views, while giving access to green, open-air terraces along the entire roofline, its profile is a carefully constructed datum line created to reflect China's flowing mountains as they rise above the dense fog below. An arched

中国很多新城的发展往往依托于廉价的标准化住宅，以保证快速获取投资回报。在中国的沿海城市北海，这种趋势更为显著，考虑到其优越的地理位置，这就显得很可惜了。面对一片长达800 m的滨水区域，假山大厦的设计师提出："可不可能在这里建造一个经济可行，高密度，而且在建筑上富有新意的住房项目呢？"

尽管名字听起来有点戏谑，假山大厦项目却呈现出高层建筑与海滨公园相融合、高层住宅与混合型住宅区相融合的真实体验。项目充分挖掘优势，为住户展现海景。此外，整个屋顶都设计有植被丰茂的露天阳台，让住户享有绿色。它的外轮廓经过了精心设计，旨在展示连绵的山峰在浓雾中升起的审美趣味。拱形入口和建筑上的开口孔洞为海滨和楼后的山谷提供了视觉和空间上的连续性，这也为设计增加了一些独特的元素。这个灵巧的设计一方面让整座建筑物产生一种宏伟的视觉效果，另一方面也打破了其隔断性，让它看起来更加通透。该方案的几何形态将两种常见但又截然相反的建筑类型——高层和低层汇合在一起，产生了起伏有致的设计效果，恍如一座山，这样的几

**Completion Date:** December 2014
**Height:** 121 m (397 ft)
**Stories:** 33
**Area:** 492,369 sq m (5,299,816 sq ft)
**Use:** Residential
**Developer:** Beihai Xinpinguangyang Real Estate Development Co. Ltd
**Architect:** MAD Architects
**Structural Engineer:** Jiang Architects and Engineers
**MEP Engineer:** Jiang Architects and Engineers

竣工时间：2014 年 12 月
高度：121 m (397 ft)
层数：33
面积：492 369 m² (5 299 816 ft²)
主要功能：住宅
开发商：北海馨平广洋房地产开发有限公司
建筑设计：MAD 建筑事务所
结构设计：上海江欢成建筑设计有限公司
机电设计：上海江欢成建筑设计有限公司

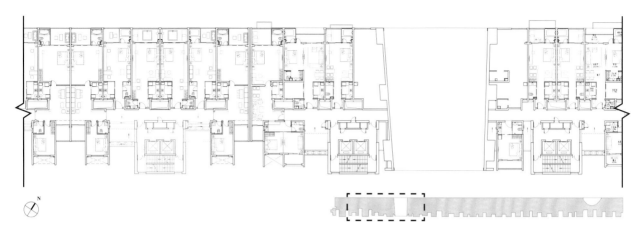

entryway and cut-out hole provide visual and spatial continuity between the oceanfront and the valley behind the building, as well as an element of drama on the inclined approach. These deft gestures simultaneously create an awareness of the building's monumentality while breaking it down and making it seem penetrable. The fundamental geometry of the scheme combines two common yet opposite architecture typologies; the high-rise and the "groundscraper," producing an undulating building typology, resulting in the form of a hill. The geometry of the architecture maximizes potential views for the residents, while a continuous platform along the ground-floor supports the public spaces, with gardens, tennis courts and swimming pools on top of the man-made hills.

Fake Hills provides a heightened experience of the coastline and an opportunity for unhindered interaction with the city and the vast body of water it faces. At its base, there are smaller pavilions overlooking a large green surface that acts as a blanket, covering an entire ground floor of programmed space, which currently includes activities as diverse as a karaoke bar, restaurant, and hotel. This green surface facilitates a landscape overlooking the gulf, and a transitional environment between beach and building, rather than pushing urban sprawl to the sea's edge.

Along with the ground-floor green roof, there is a green "strip" that runs the entire length of the top

## Jury Statement | 评委会评语

Fake Hills is the rare project that is both monumental and playful at the same time. It speaks to genuine human needs, not just the well recognized physiological need to be close to water, but also under-appreciated needs, such as irony.

在这个沿海社区，似乎到处都充斥着高层建筑，假山大厦项目的设计师则做了一些新的尝试。这个建筑罕见地集宏大和趣味于一身。它要满足的是人类真正的需求，不仅满足大家通常认为的靠近水边、感受流动的空气之类的一般生理需求，还试图满足人们对新鲜有趣事物的情感需要。

> *"All of the emotive force of a city wall or a mountain range – solidity, inscrutability, majesty – is found here, while still providing a means of escape and release through punched openings."*
>
> *"城墙或山脉所有的情感力都能在这里体现出来——坚固、神秘、雄伟,同时又通过孔道开口提供了逃离和释放的通道。"*
>
> Design Juror　设计评委

of the building. This allows for outdoor activities and recreation to take place throughout the rooftop terraces. The green roof and green strip also provide insulation to the building, reducing the need for constant temperature mitigation. By maximizing the façade facing the oceanfront, the residential units not only receive maximal sunlight exposure, but also maximal natural ventilation.

何结构可以最大限度地扩展住户的视野。在假山大厦的顶部是连通的平台,形成了一个包含着花园、网球场和游泳池的公共空间。

假山大厦让住户充分享受海岸线的景观,拥抱广阔的海面,同时和城市生活无缝互动。其底层设有小凉亭,住户们可以饱览绿草如茵的地面景观。该项目的规划土地全部被绿地覆盖,底层设有卡拉OK酒吧、餐馆和酒店等种类繁多的娱乐设施。这片绿地可以俯瞰海湾,是海滩和建筑的过渡地带,同时也避免了将城市扩张到大海的边缘。

沿着地面层的绿色屋顶,绿色飘带延展铺满了建筑物的整个顶部。这个设计可以让住户在整个屋顶露台上进行户外活动和娱乐。绿色屋顶和绿化带也成为建筑物的保护层,减少了恒温调节的需要。通过最大程度地扩大正面面对海滨的面积,各个住宅单元能够获得最佳的阳光照射与自然通风。

**Previous Spread | 前页**
Left: Overall view
左图: 全景

**Current Spread | 本页**
**Opposite Top:** Partial typical floor plan with key
对侧上图: 标准层部分平面图
**Opposite Bottom:** Pedestrian passageway providing waterfront access
对侧下图: 临水的步行通道
**Left:** Exterior view of surrounding green space
左图: 户外绿色景观

# China Best Tall Building – Honorable Distinction
中国高层建筑荣誉奖

## Hua Nan Bank Headquarters
华南银行总部大楼

Taipei | 台北

The genesis of the Hua Nan Bank Headquarters building was a request from the client for something "conservative" that lets clients see "stability" in its corporate architecture, into which a surprisingly large number of customers is expected to visit regularly — a brief that could alternately be seen as calling for a higher degree of exuberance in the design. There were also stringent restrictions on the orientation of the building on the site. These factors provoked several inventive proportional sleights-of-hand, the result of which is an elegant and well-detailed tower.

The Hua Nan Bank Headquarters is located in the busy Xinyi District of Taipei, among a cluster of other financial institutions. The building is comprised of a three-story podium with a 27-story office tower above. The main entrance and long side of the rectilinear plan face west. The pattern of the tower's façade visually merges two floors together, creating a distinctive exterior design element – a "packet" that reads as one from the outside – that

**Completion Date:** January 2014
**Height:** 155 m (507 ft)
**Stories:** 27
**Area:** 40,700 sq m (438,091 sq ft)
**Use:** Office
**Owner/Developer:** Hua Nan Commercial Bank Ltd.
**Architect:** KRIS YAO | ARTECH (design)
**Structural Engineer:** King-Le Chang & Associates
**MEP Engineer:** WSP | Parsons Brinckerhoff
**Main Contractor:** Li Jin Engineering Co., Ltd.
**Other Consultants:** Chroma33 Architectural Lighting Design (lighting); Magnificence Interiors Inc. (interior); Ming Shen Engineering Inc. (façade); PLACEMEDIA (landscape); Shen Milsom Wilke, Inc. (acoustic)

华南银行总部大楼的客户要求建筑风格"保守",希望可以让银行的顾客从建筑风格中感受到"稳定"。因为有大量的顾客会定期造访,这也对建筑容纳人流的能力提出了更高要求。项目对建筑物的朝向也有严格的限制。由于这些因素的存在,设计者通过采用多种富有创造力的巧妙手法,最终完成了这一栋风格优雅、细节处理得当的大楼。

华南银行总部大楼位于繁华的台北信义区,区内金融机构林立。整座建筑由一座3层裙房与其上27层的办公楼组成。主入口和直线形的建筑长边面朝西。该大楼的外墙设计将两层高的空间在视觉上结合在一起,从而创造出一个独特的外观设计元素——"外部结构框架"。"外部结构框架"使得大楼从外部看起来融为一体,并显得比实际更高更细。

外围结构与外部幕墙相分离,向外突出,形成了外骨骼支撑系统。在这里,铝制百叶采用对角设计,融入了中国传统的设计图案。此举有双重优点,一方面为露台创造了空间,避免阳光直射,另一方面还创造出了无柱的内部

竣工时间:2014年1月
高度:155 m (507 ft)
层数:27
面积:40 700 m² (438 091 ft²)
主要功能:办公
业主/开发商:华南商业银行股份有限公司
建筑设计:姚仁喜 | 大元建筑工场
结构设计:杰联国际工程顾问有限公司
机电设计:美国柏诚国际股份有限公司台湾分公司
总承包商:利晋工程股份有限公司
其他顾问方:大公照明设计顾问有限公司(灯光);禾安室内装修设计工程股份有限公司(室内设计);茗生工程股份有限公司(外墙设计);PLACEMEDIA(景观设计);声美华顾问公司(声学设计)

makes the tower appear taller and more slender than it actually is.

Detached from the exterior curtain wall, the peripheral structure is extruded, forming an exoskeleton support system. Here, aluminum louvers create a diagonal pattern that evokes traditional Chinese designs. This design achieves the twofold benefit of creating spaces for balconies – which also provide shade from the sun – and allowing column-free interior space. The green space at the foot of the tower features a landscape composed of native plants; its irrigation demands are fully covered by rainwater and greywater. Similar green features are found throughout the building in double-height atrium gardens.

Contrary to most conventions of sustainable design, the planning scheme for the main business district in Taipei has most of its lots facing east and west. In a culture fixated on having significant building façades towards the "front", which would be east or west in

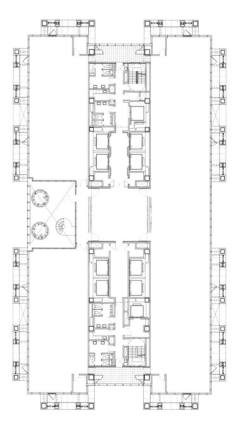

**Previous Spread | 前页**
Left: Overall view
左图: 全景

**Current Spread | 本页**
Opposite Top: Interior view of double-height space with greenery
对侧上图: 两层高配有绿景的室内空间
Opposite Bottom Left: Close-up façade view highlighting garden atria
对侧左下图: 室内绿色庭院使立面别具特色
Opposite Bottom Right: Typical floor plan
对侧下右图: 标准层平面图

## Jury Statement | 评委会评语

Customer focus today extends beyond the delivery of services to an expression of an attitude toward the community. The Hua Nan Bank Headquarters, through use of pleasing patterns, extensive greenery within and without, and welcoming spaces, announces the attitude of its owner with clarity and determination.

"以顾客为中心"的口号现在已经从提供服务发展为向社区展示一种态度。通过使用令人赏心悦目的图案、建筑内外布置大量繁茂的绿植，加上友好的空间布局，华南银行总部明确而坚定地宣示着业主的态度。

this case, this posed great challenges in designing a green building. Because the main entrance and long side of the rectilinear plan face west, solar protection became a paramount concern. Detached from the exterior curtain wall, the peripheral structure is pushed out as an exoskeleton, allowing spaces in between for balconies, which also act as sun-shading devices. This design feature allows the interior space to be column-free, with a clean and smooth window line surrounding it, allowing for flexibility in space planning. On the front center bay of the tower, a series of double-height "sky gardens" are stacked all the way to the top of the tower. They make for casual meeting and relaxing spaces for the office staff, and serve as the buffer zones for reducing heat gain from the west façade. Over-sized ceiling fans are installed in these spaces.

空间。大楼底部的绿地是由本土植物形成的特色景观，只需雨水和中水就可以满足灌溉需求。在两层挑空的中庭花园内也有类似的绿色景观。

与大多数可持续设计的惯例相反，台北主要商业区的规划中有许多东西朝向的建筑。在一个强调重要建筑必须朝"前"的文化环境中，不论是面向东方还是西方都将给建筑设计的环保节能带来巨大的挑战。因为主入口和直线形建筑物的长边面西，遮阳隔热就成为首要问题。因此其外围结构与外部幕墙相分离，向外突出，形成了外骨骼支撑系统，让结构柱中间空出来的空间成为阳台，同时也可以遮阳。此设计允许内部空间采用无柱结构，加上窗户的设计清晰简洁，为室内空间规划提供了足够的灵活性。在大厦前部，一系列两层挑空的"空中花园"堆叠直到塔顶。这些地方为办公室职员提供了一个举行非正式会面和休闲放松的区域，同时也可以作为缓冲区，以减少从西外墙吸收的热量。此外，在这些空间中还安装了超大型的微气流风扇。

*"The contradiction between rectilinear motifs during the day and circular stairs in the sky gardens, visible mostly at night, embodies the bank's obligation to seem solid and inviting at the same time."*

*"白天看到的是直线形图案，而到了夜间，空中花园的环形台阶就会显现出来，这种双重效果同时体现着银行坚定的责任感与强烈的热情。"*

Design Juror    设计评委

# China Best Tall Building – Honorable Distinction
# 中国高层建筑荣誉奖

## Lujiazui Century Financial Plaza
## 陆家嘴世纪金融广场

Shanghai ｜ 上海

This project is a large-scale office block development located in the Pudong district of Shanghai, four kilometers southeast of the famous Lujiazui skyscraper area. The plan clusters five office towers and one commercial building on a site measuring 270 meters north to south and 200 meters east to west along Yanggao Road, one of the main streets of Pudong.

Second-floor-level pedestrian decks connect the individual buildings functionally and in terms of traffic flow, and park-like landscaping provides restful spaces at ground level. Each part of the above-ground floors features distinctive designs and styles that link the spaces, while endowing the whole plan with diversity and continuity.

The defining design concept is "a city space created by 4,500 individual molecules." The idea was to create high-density urban space using minute pieces, in the same way that matter is composed in nature, by massive numbers of molecules clustered and crystallized together.

这座大型办公楼位于上海浦东，在著名的陆家嘴摩天大厦区域东南方向的 4 km 处。该项目计划包含五座办公楼和一座商业大厦，南北占地 270 m，东西占地 200 m，位于浦东的主干道杨高路上。

二楼的步行天桥从功能和人流组织上将各幢建筑物连接了起来。底层公园般的景观提供了休憩的空间。地面以上各楼层都有着特色鲜明的设计风格，既将整个空间连成一体，也赋予了整个方案以多样性和连续性。

项目最关键的设计概念是"4 500 个分子构成的城市空间"。在自然界，物质由大量分子聚集而成，在这处高密度的城市空间也同样由一个个细小的分子组成。

如今的中国城市高楼林立，人类与之相比仿佛沧海一粟。项目的设计师们试图寻求一种将人与建筑、建筑与城市相联结的模式。研究小组分析了门廊的图案、石库门建筑风格以及其他中国传统建筑空间的特点，据此设计了占地面积为 3 m×9 m 的双层单元"部件"（分子）。

**Completion Date:** June 2014
**Height:** Tower 1: 190 m (623 ft); Tower 2: 130 m (427 ft); Tower 3: 130 m (427 ft); Tower 5: 80 m (262 ft); Tower 6: 80 m (262 ft)
**Stories:** Tower 1: 41; Tower 2: 28; Tower 3: 28; Tower 5: 17; Tower 6: 17
**Area:** Tower 1: 128,580 sq m (1,384,024 sq ft); Tower 2: 96,730 sq m (1,041,193 sq ft); Tower 3: 93,490 sq m (1,006,318 sq ft); Tower 5: 57,780 sq m (621,939 sq ft); Tower 6: 58,390 sq m (628,505 sq ft)
**Use:** Office
**Owner/Developer:** Shanghai Lujiazui Financial and Trade Zone Development Co., Ltd.
**Architect:** Nikken Sekkei Ltd.; CCDI Group; HaiPo Architects
**Structural Engineer:** Nikken Sekkei Ltd.; CCDI Group; HaiPo Architects
**MEP Engineer:** Nikken Sekkei Ltd.; CCDI Group; HaiPo Architects
**Main Contractor:** Shanghai Installation Engineering Co., Ltd.; Shanghai Construction No. 1 (Group) Co., Ltd.

竣工时间：2014 年 6 月
高度：塔 1：190 m (623 ft)；塔 2：130 m (427 ft)；塔 3：130 m（427 ft）；塔 5：80 m (262 ft)；塔 6：80 m (262 ft)
层数：塔 1：41；塔 2：28；塔 3：28；塔 5：17；塔 6：17
面积：塔 1：128 580 m² (1 384 024 ft²)；塔 2：96 730 m² (1 041 193 ft²)；塔 3：93 490 m² (1 006 318 ft²)；塔 5：57 780 m² (621 939 ft²)；塔 6：58 390 m² (628 505 ft²)
主要功能：办公
业主／开发商：上海陆家嘴金融贸易区开发有限公司
建筑设计：株式会社日建设计；CCDI 悉地国际；HPA 海波建筑设计
结构设计：株式会社日建设计；CCDI 悉地国际；HPA 海波建筑设计
机电设计：株式会社日建设计；CCDI 悉地国际；HPA 海波建筑设计
总承包商：上海市安装工程集团有限责任公司；上海建工一建集团有限公司

China's cities today are filled with high-rise buildings that defy human scale. For this project, the designers sought a scale that would connect human beings to the architecture and the architecture to the city. The team analyzed motifs of doorways, shikumen stone-gate-style architecture and other features found in traditional Chinese architectural spaces, and based the exterior design on double-floor, 3- x 9-meter units as the "pieces" (molecules).

The internal urban space borne out of some 4,500 such pieces produces, within the five mammoth volumes of these buildings, spaces that are suitable to human scale. In the course of the design, responding to the client's request for a unified and dignified exterior look, the team proposed a deeply articulated and finely shaped façade that would shield against strong sunlight, enhancing the comfort of the interior space.
In order to identify problems and raise the development and technical level of the whole in the course of the construction, a unified design system for both the exterior and interior of the buildings was adopted.

The project incorporates open, heavily planted landscaping that helps to cool down the ground surface, contributing to the urban environment by mitigating the heat-island phenomenon. Also, the integrated use of the underground concrete frame as a district-wide block, and the introduction of advanced engineering methods, allowed a reduction in the quantity of reinforced concrete retaining walls and struts needed for the excavation, in turn allowing for a major reduction in the amount of materials needed and collateral waste generated in the construction, as well as shortening construction time.

The curtain walls use deeply articulated eaves that shield the space from the sun. Installation of natural ventilation vents helps reduce the amount of power needed for air conditioning during the moderate-weather months in spring and autumn. The design reduces glare and heat reflection by up to 60% compared to standard glass.

> **"The attention paid to the human scale, in an area that sorely needs it, is one of the defining high-quality characteristics of the Lujiazui Century Financial Plaza."**

"对人性化的关注是一直以来迫切需要的，这也是陆家嘴世纪金融广场最典型的高品质特色之一。"

Design Juror　设计评委

## Jury Statement | 评委会评语

The architecture of the project reinvigorates the disciplined study of proportion and composition in the modern Chinese skyscraper, even as many forces conspire to cast such discipline aside in the name of expediency. The project team is to be commended for persevering in the cause of lasting quality.

　　陆家嘴世纪金融广场的建造对摩天大楼的比例和构成进行了严谨的研究。该项目团队坚持项目建设的高质量，非常值得称誉。

**Previous Spread | 前页**
Left: Overall view at night
左图: 夜景

**Current Spread | 本页**
Opposite Left: Low-level typical floor plans – Tower 1 (top), Towers 2 & 3 (middle), Towers 5 & 6 (bottom)
对侧左图: 低层区标准平面图——塔1（上图），塔2（中图），塔5、塔6（下图）
Opzposite Right: Detail view of façade modules
对侧右图: 立面细部

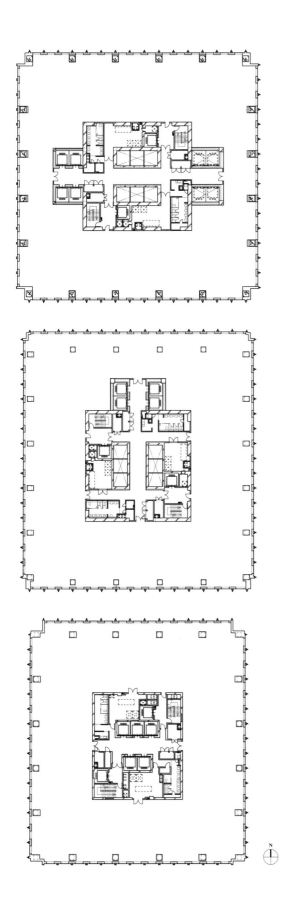

这座城市空间的内部就是由大约 4 500 个这样的部件组成的，旨在这五栋庞大的建筑物中创建出与人类尺度相符的空间。在设计的过程中，为了满足客户对外部设计兼具统一与尊贵的需求，设计团队提出了一个结构明晰而美观的外墙设计方案，一方面可以阻挡强日光照射，另一方面又提高了内部空间的舒适度。

为了确定施工过程中遇到的问题，提高施工过程的整体开发和技术水平，建筑外部和内部采用了统一设计的体系。

该项目包含一片高植被覆盖的露天景观区，这有助于降低地面温度，保护城市环境，缓解热岛效应。此外，通过综合使用与街区等宽的地下混凝土结构，在引进先进的工程方法的基础上，钢筋混凝土挡墙的数量与挖方所需支撑的数量都得以减少。如此一来，建筑过程中原料的需求量与附属废品的生成量也随之减少，施工时间得以大大缩短。

为了抵挡阳光，幕墙使用了经过精密计算的屋檐结构。自然通风孔有助于减少在温和的春季和秋季对空调的需求。相比一般的玻璃，该设计减少了高达 60% 的光热反射。

37

# China Best Tall Building – Honorable Distinction
中国高层建筑荣誉奖

## Nanchang Greenland Central Plaza
南昌绿地中心

Nanchang | 南昌

Twin towers define the Jiangxi Nanchang Greenland Central Plaza, Parcel A project. The identical towers, 100 meters apart at the center, anchor an emerging business, civic, and residential zone in one of Nanchang's booming new areas, and represent the only curving glass towers in the city.

The twins' shapes morph from base to top: at ground level, the towers are square in plan in order to provide optimal site access, while at their crowns, the square has been rotated 45 degrees, so as to align the buildings with prime views of the Ganjiang River. Throughout its height, the square is beveled with smooth rounded corners, further emphasizing the views and perspective shifts from floor to floor as the plan rotates.

Inside, both towers offer 59 stories of leasable office space anchored by a transparent ground-floor lobby. At each level, the geometry of the structural cores

江西南昌绿地中心 A 地块包括两座塔楼。两座塔楼完全一样，中间相隔 100 m，它们标示着在蓬勃发展的南昌新城中，一个新兴的商业中心、市民活动和居住区正日渐成型。这也是南昌唯一的曲面玻璃建筑项目。

双塔的形状从基部到顶部具有丰富的动态变化：在地面层，大厦平面是方形的，最大限度地实现了交通通达性；到了建筑顶端，平面旋转了 45°，以获得观赏赣江的最佳角度。整座建筑的方形楼层都配有平滑的圆角，使不同楼层都有不同的景观视角。

双塔内部均包括 59 层可租用的办公空间以及一个透明的底层大厅。每层结构核心筒的几何形状都与每层建筑平面形状保持一致，这样可以确保使用空间的进深在各边都保持一致。即使在两座大厦随着高度的变化而转变角度的过程中，这一点也得到了保证。恒定的进深空间也确保了最佳的视野。

双塔通过高效利用材料和风力管理达到了可持续发展的目标。运用模块化、冷弯板的系统设计，大厦呈现出波

**Completion Date:** January 2015
**Height:** Tower 1: 303 m (994 ft); Tower 2: 303 m (994 ft)
**Stories:** Tower 1: 59; Tower 2: 59
**Area:** Tower 1: 110,000 sq m (1,184,030 sq ft); Tower 2: 110,000 sq m (1,184,030 sq ft)
**Use:** Office
**Developer:** Greenland Group
**Architect:** Skidmore, Owings & Merrill LLP; ECADI
**Structural Engineer:** Skidmore, Owings & Merrill LLP
**MEP Engineer:** Skidmore, Owings & Merrill LLP
**Other Consultants:** Aon Fire Protection Engineering (fire); Edgett Willams Consulting Group Inc. (vertical transportation); Kaplan Gehring McCarrol Architectural Lighting, Inc (lighting); Lerch Bates (façade); RWDI (wind); Shen Milsom Wilke, Inc. (acoustics); SWA Group (landscape)

竣工时间：2015 年 1 月
高度：塔 1: 303 m (994 ft)；塔 2: 303 m (994 ft)
层数：塔 1: 59；塔 2: 59
面积：塔 1: 110 000 m² (1 184 030 ft²)；塔 2: 110 000 m² (1 184 030 ft²)
主要功能：办公
开发商：绿地集团
建筑设计：Skidmore, Owings & Merrill LLP；华东建筑设计研究总院
结构设计：Skidmore, Owings & Merrill LLP
机电设计：Skidmore, Owings & Merrill LLP
其他顾问方：Aon 消防工程公司（防火）；Edgett Willams 咨询集团（垂直交通）；Kaplan Gehring McCarrol 建筑照明设计公司（灯光）；Lerch Bates 顾问公司（外墙）；RWDI（风工程）；声美华顾问公司（声学）；SWA 集团（景观）

> *"The twin towers achieve a remarkable synthesis of aesthetic appeal and uncompromising aerodynamic engineering. Their translucent character is the asset that will allow them to remain landmarks for a long time."*
>
> "双子塔实现了审美情趣与空气动力学的完美融合。其晶莹剔透的外形使它们能够长期保持地标建筑的地位。"

Design Juror　设计评委

ensures that lease depths remain constant on all sides. This remains true even as each tower changes shape over the course of its height. The clear lease depths also allow for optimal views.

The twin towers address sustainability through the efficient use of materials and wind management. The building skin was designed using systems of modularized, cold-bent panels. At the top of each tower, the design team precisely spaced and angled standard glass panels to create porous crowns. The crown design maximizes porosity in the prevailing wind direction in order to minimize wind loading at the base of the towers.

The crown also represents a unique solution to a challenge faced after construction had begun on the tower: the design team received a request to increase height from 289 to 303 meters. As additional wind load at the top of a tower has a significant impact on the forces at the base, this seemingly small increase would have generated a large impact on the structure. To alleviate this issue, the design team modified the tower so that wind loads on the taller tower would be reduced. After investigating many shape modification and porosity options, the final "porous crown" solution was selected. When compared to a non-porous solution, it reduces the total wind load on the tower by 12%, thus achieving the projected wind load levels of the original 289-meter design. In addition to being an effective anti-drag device, it is also an inspiring space to occupy, allowing just enough exposure to the elements to communicate the thrill of standing at this height, while still offering a feeling of enclosure and safety.

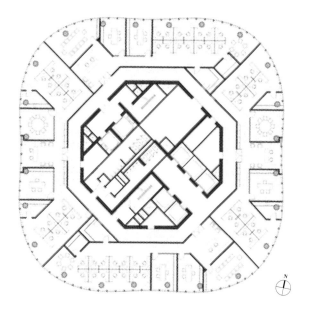

浪形的轮廓。这一主题在细节之处的设计也多次体现,如大堂中从核心筒向外发光的照明板。在两座大厦的顶部,设计团队通过精心定位标准化玻璃板,创造出了透空的冠顶。冠顶的设计最大限度地提高了主导风向上的孔隙率,从而减少了风荷载对大楼底部的压力。

这顶皇冠也是设计团队为应对一项特殊挑战提出的解决方案。开工后,设计团队收到了一个任务,要求把大楼的高度从 289 m 增加到 303 m。由于大楼顶部额外的风荷载会对塔基有显著影响,这看似微小的高度增加,实际上对已建成的结构会产生很大的影响。为了缓解这一影响,设计团队从空气动力学的角度出发,改进了塔楼结构,以减少高度增加为大楼带来的额外风荷载。在研究了许多种形态方案和孔隙方案之后,设计团队最终选择了"多孔冠"这一解决方案。相比于非多孔方案,这一设计可以减少 12% 的风荷载,从而将预期的风荷载水平降低到原始 289 m 设计的总风荷载量。它不仅是有效的抗阻装置,更是一片令人神清气爽的空间。站在这里,你既能感受到高空中各种元素带来的刺激感,同时也不乏封闭和安全感。

## Jury Statement | 评委会评语

The Jiangxi Nanchang Greenland Central Plaza towers may superficially resemble a pair of chimneys, but they represent the future white-collar workplace, speaking of transparency as much as efficiency. They represent China's transition from smokestack industry into a multivalent economic powerhouse.

虽然江西南昌绿地中心看起来可能像一对工业烟囱,但它们实际上是未来的白领工作场所,注重采光、效率、活力和一致性。它们代表了中国正在从一个浓烟工业国家向多元化经济强国的转变。

**Previous Spread | 前页**

Left: Overall view in context
左图: 全景

**Current Spread | 本页**

**Opposite Bottom:** View showing details in the crown
对侧下图: 冠顶细部
**Top Left:** Detail view of the crown
上左图: 冠顶细部
**Top Right:** Typical floor plan
上右图: 标准层平面图

# China Best Tall Building – Honorable Distinction
中国高层建筑荣誉奖

## Nanchang Greenland Zifeng Tower
南昌绿地紫峰大厦

Nanchang | 南昌

Like many secondary cities in China, Nanchang is currently undergoing rapid development from the ground up, with aesthetics taking a backseat to speed. Its Gaoxin district is very much under construction, with neutrally colored and conventionally shaped high-rise complexes characterizing the area. By contrast, the Jiangxi Greenland Zifeng Tower is a striking presence with its deeply carved shape and reflective facade.

Office floors occupy the bottom two-thirds of the tower. The regular rectangular form and a central core ensure that these floors remain incredibly flexible. A luxury hotel stacks on top of the office floors, occupying the tower's top 18 stories. This change in program is registered by the Great Window—a distinct aperture carved into the tower's mass. Formally, the Great Window transitions the floor plates to a suitable size and layout for the hotel function. Aesthetically, it marks the tower with a strong design, distinguishing it from the other towers that comprise Nanchang's skyline.

和许多中国二线城市一样，南昌目前正处于快速发展期，审美需求也搭上了顺风车加速起飞。目前南昌高新区有许多项目正在施工中，大多是色调柔和的常规高层建筑。相比之下，南昌绿地紫峰大厦以其雕塑般的艺术造型和闪亮的色彩别具一格。

办公区占据了大厦底部三分之二的空间。每个楼层采用长方形加核心区的平面，确保每层的布局都有足够的灵活度。办公区以上的18层是一所豪华酒店。功能的转变在"城市之窗"中体现出来，这是一个嵌入在塔楼体量的标志性特色。从形式上看，"城市之窗"使楼面过渡到适合酒店功能的尺寸和布局。从美学上说，它使塔楼恢弘大气，也使建筑在构成南昌城市肌理的塔楼中脱颖而出。

大厦采用玻璃幕墙的设计，将这一建筑从周围环境中凸显出来。根据光照强度，玻璃幕墙会折射出绚丽的颜色。斜肋构架的遮阳板设计也给大厦增添了独特的几何质感，使大厦底部的三分之二与顶部的"城市之窗"形成强烈反差。

**Completion Date:** August 2015
**Height:** 268 m (879 ft)
**Stories:** 56
**Area:** 209,058 sq m (2,250,282 sq ft)
**Use:** Hotel / Office
**Developer:** Greenland Group
**Architect:** Skidmore, Owings & Merrill LLP; ECADI
**Structural Engineer:** Skidmore, Owings & Merrill LLP
**MEP Engineer:** Skidmore, Owings & Merrill LLP
**Other Consultants:** Edgett Willams Consulting Group Inc. (vertical transportation); Kaplan Gehring McCarrol Architectural Lighting, Inc (lighting); Lerch Bates (façade maintenance); Rolf Jensen & Associates (fire); RWDI (wind); Shen Milsom Wilke, Inc. (acoustics); SWA Group (landscape)

竣工时间：2015年8月
高度：268 m (879 ft)
层数：56
面积：209 058 m² (2 250 282 ft²)
主要功能：酒店／办公
开发商：绿地集团
建筑设计：Skidmore，Owings & Merrill LLP；华东建筑设计研究总院
结构设计：Skidmore，Owings & Merrill LLP
机电设计：Skidmore，Owings & Merrill LLP
其他顾问方：Edgett Willams 咨询集团（垂直交通）；Kaplan Gehring McCarrol 建筑照明设计公司（照明）；Lerch Bates 顾问公司（幕墙维护）；罗尔夫杰森消防技术咨询有限公司（消防）；RWDI（风工程）；声美华顾问公司（声学）；SWA 集团（景观）

The Jiangxi Greenland Zifeng Tower glass façade also delineates the building from its surroundings, refracting dynamic colors that are determined by the quality of natural light. The diagrid shading fin system adds a unique, strictly geometric texture to the building, contrasting the lower two-thirds of the tower with the Great Window at the top.

Functionally, the tower brings much-needed luxury hotel, retail, and office space to Gaoxin. Many of these activities are accessible through the attached podium building at ground level, helping to open up the project site as a public thoroughfare in the neighborhood. The ground plane has also been landscaped so as to create a cohesive design that extends from the ground plane to the building façade and podium rooftop, encouraging people to pass through the site and enter the tower and podium buildings. This, in turn, creates a property of high value for the client, helping to satisfy overarching business goals.

The system of perforated, aluminum triangular fins that trace up all four elevations anchors the building's high performance features. The fins shade the tower; preventing excessive solar heat gain and glare to the interior. The design team laid the fins out in a diagrid pattern that wraps the entire building, so as to maximize these high-performance benefits. The fins are also programmed with tiny, white high-efficiency LED lights. When illuminated at night, the LEDs serve as the tower's primary exterior lighting scheme, reducing energy loads typically associated with this design element.

在功能上，大厦为高新区带来了发展急需的豪华酒店、购物区和办公空间。大厦的很多功能区都可以通过一层的裙房进入，裙房的设置也使大厦所在地块变得开放，从邻里街区可以畅通穿行于此。建筑所在场地经过了专门的景观设计，力图使底层、外墙和裙房屋顶保持统一的设计感，吸引人们从这里进入大厦和裙房。这些设计反过来为客户创造了很高的价值，帮助满足其总体经营目标。

绿地紫峰大厦建筑四面使用了三角形铝材遮阳板，有助遮阳，还可防止过多的太阳能辐射和内部炫光。设计团队按照斜肋构架让遮阳板包裹了整栋建筑，从而最大限度地发挥其效用。遮阳板还自带一体化小型高效白色 LED 照明，这些 LED 是该建筑主要的夜间外观照明设备，能有效降低夜间外观照明能源负荷。

**Previous Spread | 前页**

**Left:** Overall view in context
**左图:** 城市环境中的建筑全景

**Current Spread | 本页**

**Bottom:** Building entry
**下图:** 建筑入口

**Opposite Left:** Building section
**对侧左图:** 建筑剖面图

**Opposite Right:** Overall view of the illuminated building
**对侧右图:** 建筑灯光效果全景

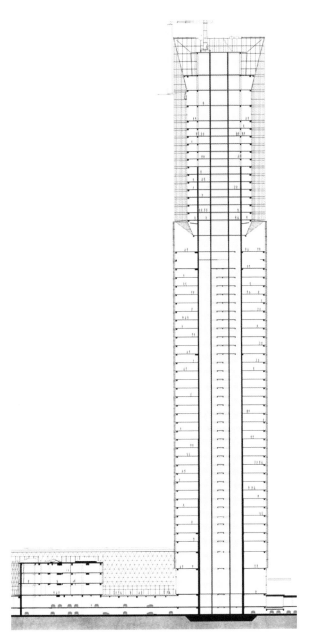

### Jury Statement ｜ 评委会评语

At Jiangxi Nanchang Greenland Zifeng Tower, a significant portion of the site is given over to public access and a park-like atmosphere, something that is only possible through a skilled execution of a tall building on the site. The quality of the ground-level space matches that of the thoughtful building design.

江西南昌绿地紫峰大厦非常重视向公众开放，并且营造出公园般的氛围，对于高层建筑来说，只有通过高超的设计技巧才能达到这两个目标。景观设计质量与整体建筑的设计也非常匹配。

*"The fractal-like, beveled façade of the Jiangxi Nanchang Greenland Zifeng Tower makes a strong statement of personality for a new-build district, giving it a distinguished presence on the skyline."*

"江西南昌绿地紫峰大厦如碎片般带斜角的立面强烈表现出这个新建筑区的发展势头，而其本身也成为天际线上一道靓丽的风景线。"

Design Juror　设计评委

# China Best Tall Building – Honorable Distinction
# 中国高层建筑荣誉奖

## OLIV | 香港OLIV
Hong Kong | 香港

OLIV is a retail building within the busy Causeway Bay shopping district in Hong Kong, near other retail landmarks such as Hysan Place, Leighton Center, and Times Square. Sitting on a small site of just over 280 square meters, the narrow building is nonetheless able to offer a floor-to-floor height of 5 meters on each floor, with one large unit per floor. Without passing through lobbies or corridors, occupants are delivered by elevator directly into unobstructed tenant spaces, each with columns only at the corners. OLIV is a product of a "total design" theory that encompasses the smallest details in pursuit of contemporaneousness.

The design is inspired by the olive tree, after which the building was named. The floor plate changes slightly at each level, as the building moves upward, giving it a twisting and turning profile. The exterior is wrapped with a "triple skin" on the outside, against an interior orthogonal layer of subframe and mullions, a dark grey glass curtain-wall layer, and an organic exterior

**Completion Date:** January 2014
**Height:** 136 m (445 ft)
**Stories:** 25
**Area:** 4,305 sq m (46,339 sq ft)
**Use:** Retail
**Owner/Developer:** Benway Limited
**Architect:** THEO TEXTURE; Studio Raymond Chau Architecture Limited
**Structural Engineer:** Wong & Cheng Consultants Engineers Limited
**MEP Engineer:** Trustful Engineering & Construction Co., Ltd.
**Main Contractor:** Trustful Engineering & Construction Co., Ltd.
**Other Consultants:** Harbour Century Limited (quantity surveyor)

OLIV作为一栋商业建筑位于繁忙的铜锣湾购物区，紧邻希慎广场、礼顿中心和时代广场等地标性购物中心。虽然占地面积只有280多平方米，但在这一狭窄的建筑中，每层楼高都高达5 m，并且均为大面积的独立单元。无需通过大厅或走廊，住户便可以经由电梯畅通无阻地直接进入大楼中的租户空间。大楼的立柱被布置于每层楼的角落处，从这些设计中不难窥见OLIV"总体设计"理念，在追求共时性体验的同时对最微小的细节也绝不放过。

大厦的设计灵感来自橄榄树，该建筑的名称也由此得来。每一层的平面都随着层数的增高有着细微的变化，这使建筑物整体形成了略带扭曲的轮廓。外墙有三层：内层的直角支撑结构、深灰色的玻璃幕墙和白色金属构成的有机包层。这一包层包含的"橄榄树结"可以变成发光的"星星"，拼出了建筑的名字"OLIV"。在晚上，当小灯泡点亮之后，OLIV的标识尤其显得光彩夺目。

由于场地的大小和极具挑战性的城市环境，设计团队自从第一天接手项目之时就面临巨大的困难：包括拆走原

竣工时间：2014年1月
高度：136 m (445 ft)
层数：25
面积：4 305 m² (46 339 ft²)
主要功能：商业零售
业主 / 开发商：Benway 有限公司
建筑设计：THEO TEXTURE；周文渭建筑师事务所有限公司
结构设计：黄郑顾问工程师有限公司
机电设计：卓誉建筑工程有限公司
总承包商：卓誉建筑工程有限公司
其他顾问方：Harbour Century 有限公司（工程造价）

> *"From the smallest interior design elements to its broad, expressive exoskeleton, OLIV is redolent with creativity and stands as an exciting and intense response to a tiny urban site."*
>
> "从最小的内部设计元素到宏大且富表现力的外部构造，OLIV 充满了创造力，在香港这个玲珑之地成为一块宏伟而绚丽的地标。"
>
> Design Juror　设计评委

layer of white cladding. This layer contains the "olive tree knots" which turn into radiating "stars," as the signature and logo of OLIV, an effect particularly noticeable when lit by small point lights at night.

Due to the dimensions of the site and the challenging urban environment, the design team faced tremendous difficulties from the first day, including the demolition of the original six-story walk-up structures on site, as well as implementing the bore-pile foundation and the superstructure with its variable floor plates. The construction project took more than three years total.

The project was even more challenging from the perspective of the curtain wall design and production, in which every piece of the organic-shaped aluminum cladding was different, due to the irregular form of the building. Each piece was pre-fabricated off-site, transported to the site, and assembled to fit perfectly onto the concrete structure, forming its signature triple-layered skin.

A wind-tunnel test was conducted in order to slim down the reinforced concrete structure by about 30%, which in turn increased the efficiency of the floors. The resulting spaciousness is ideal for commercial buildings with ground floor retail and "upstairs" boutique restaurants and retailers, following a pattern that has been effectively used in Tokyo's Ginza district and has now been transposed to Hong Kong.

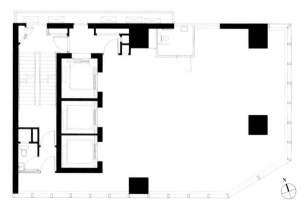

## Jury Statement ｜ 评委会评语

Many tall buildings in Hong Kong respond to its spatial scarcity by extruding a rectangular shape as high as it can go. Instead, OLIV transmits dynamism, even discomfort with this regimented approach, using its twisting form and organic references to deliver unique interior spaces and a memorable streetscape.

在香港，为了应对空间异常稀缺的限制，许多高层建筑的结构都是呈长方体形态，建筑高度则达到实际所能允许达到的最高极限。与之相反，OLIV 打破了这种刻板的做法，试图传递一个多元化的理念，它以旋转的形式和有机的表达，创造出了独特的室内空间和令人难忘的街景。

有的六层建筑，打牢钻孔桩的地基，以及处理复杂多变的楼层结构等问题。项目施工共历时三年多。幕墙设计和生产带来的挑战更大。由于不规则的建筑外形，每一块有机铝包层都有所不同。每个包层都需要提前预制，然后运到现场组装。但最终，它们和混凝土结构完美结合，构成了极具特色的三层外墙。

为使钢筋混凝土结构的厚度降低 30% 左右，从而提高每层楼的利用效率，还进行了一次风洞试验。试验证明其宽敞度非常适合一楼零售、楼上为高档餐厅和柜台的功能布置，这种大厦模式曾在东京银座区卓有成效，如今也因地制宜地应用到了香港的建筑中。

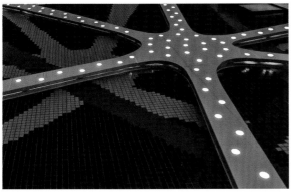

**Previous Spread ｜ 前页**

**Left:** Overall view looking up
左图：仰视全景

**Current Spread ｜ 本页**

**Opposite Bottom Left:** Interior view of sky lounge
对侧下左图：空中休息室内景

**Opposite Bottom Right:** Typical floor plan
对侧下右图：标准层平面图

**Middle:** Overall view looking up at the illuminated building
中图：建筑照明夜景

**Bottom:** Detail view of radiating "star" feature
下图：屋顶细部——闪亮的星星

# China Best Tall Building – Honorable Distinction
中国高层建筑荣誉奖

## People's Daily New Headquarters
人民日报新总部

Beijing | 北京

People's Daily is the top official newspaper in China, and the site of its new headquarters is a prominent location in the heart of Beijing's CBD, about eight kilometers east of the Forbidden City. The design was a result of a national competition held in 2009. The winning design was the result of an effort to offer a headquarters building that was creative and contemporary, but which would also reflect culture and tradition. The building also needed to be cost- and use-efficient, and environmentally friendly.

The shape of the building is an organic, "perfect" form, in which no corners are visible from any angle. The shape is defined by strict geometric calculation, as an interpretation of an old Chinese saying: "inner rectangular, outer round shape." Roughly translated, it means "one must adhere to one's principles while maintaining a modest and friendly manner."

As Beijing is a high seismic zone, the building is supported with three angled spatial truss supports,

《人民日报》是中国最高级别的官方报纸。它的新总部位于北京 CBD 的显要位置，距离故宫约 8 km。采用的是在 2009 年举行的全国竞赛胜出方案，该方案力图做到既富有创意和时代性，又能彰显文化和传统。该建筑还需要保证节约成本、高效利用和环境友好。

从外形上看，该建筑物是一个有机的"完满"结构，从任何方向上看都看不到棱角。该形状经过了严格的几何计算，正体现了古老的中国成语"外圆内方"，意思是："外表随和，内心严正。"

北京位于地震高发区，因此该建筑使用了三角空间桁架与电梯核心筒协同支撑，并将通用垂直柱和混凝土地板作为次级结构部件。

三维圆形的设计很难通过金属或玻璃板呈现，因此，设计小组继承了中国沿袭 2 500 年之久的建筑传统，使用了陶板。陶的应用远至地标性建筑秦始皇陵内数以千计的兵马俑像，近至全国各地民居楼的屋顶瓦片。人民日报总

**Completion Date:** April 2015
**Height:** 180 m (590 ft)
**Stories:** 33
**Area:** 88,856 sq m (956,438 sq ft)
**Use:** Office
**Owner/Developer:** General Office of People's Daily New Headquarters Construction & Development Program
**Architect:** Architure and Engineers Co., Ltd. of Southeast University
**Structural Engineer:** Architure and Engineers Co., Ltd. of Southeast University
**MEP Engineer:** Architure and Engineers Co., Ltd. of Southeast University
**Main Contractor:** Xinxing Construction & Development Corporation
**Other Consultants:** Beijing Institute of Architectural Design (lighting)

竣工时间：2015 年 4 月
高度：180 m (590 ft)
层高：33
面积：88 856 m$^2$ (956 438 ft$^2$)
主要功能：办公
业主 / 开发商：人民日报新总部建设和发展项目办公室
建筑设计：东南大学建筑设计研究院
结构设计：东南大学建筑设计研究院
机电设计：东南大学建筑设计研究院
总承包商：新兴建设开发总公司
其他顾问方：北京市建筑设计研究院（照明）

working in conjunction with the elevator core, while regular vertical columns and concrete floors perform as secondary structural components.

The 3-D rounded shapes would have been difficult to render in metal or glass panels. Instead, the design team used glazed terra-cotta, continuing a Chinese practice some 2,500 years old, which has been used in such landmarks as Emperor Qin's Mausoleum, with its thousands of "terra-cotta warrior" figures in Xi'an, as well as in the vernacular roof tiles found all over China. The terra-cotta "baguettes" on People's Daily Headquarters resist decay and color fading over long time periods, and their pliability into different lengths and cutting angles for their end surfaces allowed them to be woven together to achieve the delicate 3-D-curved shape.

A high-strength cable structure is used to connect these baguettes. There are more than 223,000 baguettes used throughout the structure, ranging in length from 650 to 1200 mm. There is a 900 mm space between the woven terra-cotta baguettes and aluminum sandwich insulation panel, which helps modulate interior temperatures. Air automatically flows between the double skins, providing a grey protection layer. As the designers were conscious of the poor air quality in Beijing and its likely effect on tall building surfaces, the glazed terra-cotta baguettes were also chosen because their surfaces repel dust, and what dust does accumulate washes down through the inner surface of the baguettes to feed the groundwater irrigation system. Overall, the building is more than 80% prefabricated, and is intended to be dismantled and recycled when it reaches its end of life.

Inside the building, curved corridors and public spaces sustain the visual interest of the unusually shaped exterior. Due to its rounded parallel edges, the usable floor area ratio is also highly effective at 73%, while similar office buildings in Beijing average only 53% efficiency.

## Jury Statement | 评委会评语

Society is accustomed to the idea that newspapers and other materials should be recycled. The stance taken by People's Daily, in commissioning a headquarters that could be recycled, uses the power of both media and architecture to communicate environmental messages, and envision a better future.

社会普遍认为报纸和其他材料都应该进行回收利用。人民日报也持有这个立场，它委托建造一个可以自体循环利用的总部大厦，试图利用媒体和建筑的力量来传达关键的环境态度，并构想一个比现在更美好的未来。

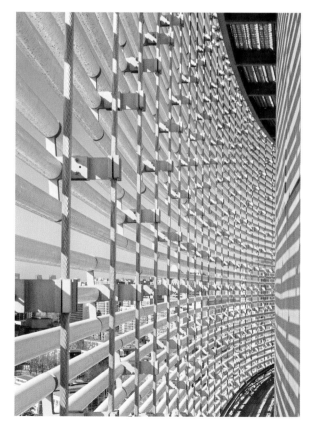

**Previous Spread | 前页**
Left: Context view showing the building's organic form
左图：全景

**Current Spread | 本页**
Opposite Left: Conference hall entry
对侧左图：会议厅入口
Opposite Right: Detail view of double façade
对侧右图：双层立面表皮的细部
Left: Building section
左图：建筑剖面

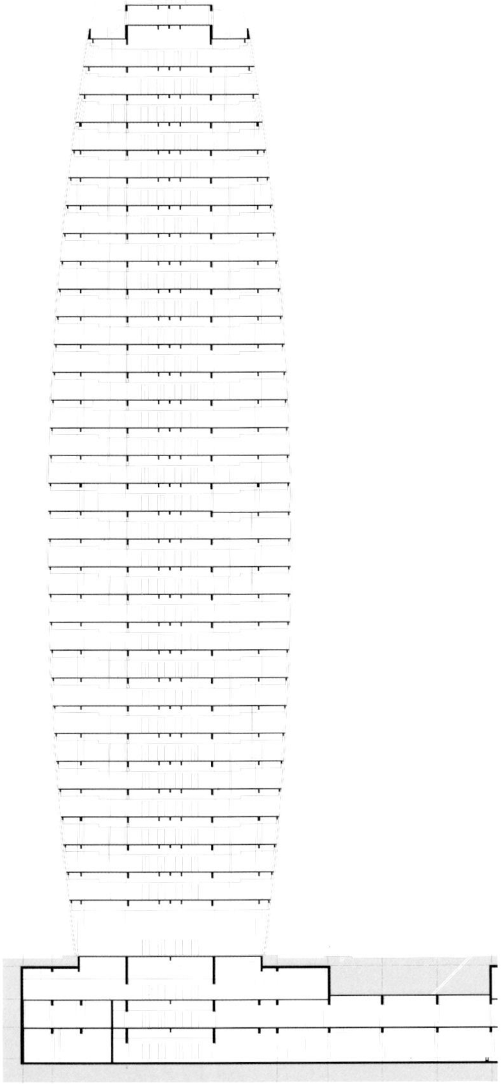

部的陶板"长棍"可保持长时间不腐坏、不褪色，同时也具有柔性，可以切割成各种长度和角度，相互交错，构建出精巧的3D曲形。

为了将这些长棍连接起来，建筑使用了高强度悬索结构。整个结构使用了超过223 000个砖块，长度从650 mm至1 200 mm不等。在编织形的陶板层和铝夹芯保温板之间有一个900 mm的空隙，作用是调节室内温度。空气在两层之间自由流动，从而提供了一个"灰色"的保护层。设计师也意识到北京不尽如人意的空气质量对高层建筑表面可能产生的影响，因此陶板上了不易附着尘土的釉质，即便蒙上厚厚的灰尘，也可以通过雨水将其冲刷下来，进入地表排水系统。总体而言，该建筑超过80%的材料都是预制的，可被拆卸分解，回收利用。

相比于形状奇特的外表，建筑内部的弯曲走廊和公共空间也不甘示弱地展示出视觉趣味。由于外墙呈弧状，可用的楼面面积比率高达73%，而类似的北京写字楼平均只有53%。

# "The People's Daily New Headquarters is an accomplished reconciliation of the most advanced contemporary technology and the requirement to create a 'classic' design through interpretation of a proverb."

"人民日报新总部是通过对某一寓意的阐释而将最先进的技术和对经典设计的需求相融合的产物。"

Design Juror　设计评委

# China Best Tall Building – Honorable Distinction
# 中国高层建筑荣誉奖

## Taiping Finance Tower | 深圳太平金融大厦
Shenzhen | 深圳

This high-rise office tower occupies a site on the northwest corner of the Shenzhen CBD Central Park. The client, one of the largest insurance companies in China, places a high priority on conveying sincerity and confidence. The design reflects this in an understated, elegant and environmentally conscious way.

The design takes the notion of the city grid and the diagonal axis oriented toward the Central Park into the façade. Outer columns and beams are expressed as triangular in section, acting as sun shading, while their slanted profiles and reflective surfaces allow occupants maximum views toward Central Park. The building has a sophisticated façade, expressing stability, dignity, and a sense of dynamism that enables transformation, providing a fitting image for the client's corporate identity.

Utilizing the double-tube system, the most popular structural system in China, the central core is replaced with an open space, with exposed elevators and

**Completion Date:** December 2014
**Height:** 228 m (748 ft)
**Stories:** 48
**Area:** 131,281 sq m (1,413,097 sq ft)
**Use:** Office
**Owner:** China Taiping Insurance Group Co.; Taiping Life Insurance Co., Ltd. Shenzhen Branch; Taiping General Insurance Co., Ltd.
**Developer:** Taiping Property(Shenzhen) Co., Ltd.
**Architect:** Nikken Sekkei Ltd.; Shenzhen General Institute of Architectural Design and Research Co., Ltd.
**Structural Engineer:** Nikken Sekkei Ltd.; Shenzhen General Institute of Architectural Design and Research Co., Ltd.
**MEP Engineer:** Nikken Sekkei Ltd; Shenzhen General Institute of Architectural Design and Research Co., Ltd.
**Main Contractor:** China Construction Third Engineering Bureau Co., Ltd.

这座超高层办公楼位于深圳福田CBD市民广场的西北角，其开发商为中国最大的保险公司之一，他们希望通过这座建筑传达出真诚与信任的重要理念。设计方案低调、典雅、环保，充分反映了这一要求。

建筑设计考虑了城市网络的构造，将垂直交汇的城市轴和面向市民广场的斜向轴应用在建筑立面设计上。比如，外围柱和梁呈三角形构造，不仅可以提供遮阳的功能，而且让建筑因视角差异呈现丰富多彩的立面效果，而倾斜的轮廓和反光表面更能让建筑内的人们最大程度地饱览市民公园的风景。建筑外观设计巧妙精细，表现出了稳定与尊贵的风格，同时传达出一种勇于革新的活力，完全符合开发商对企业形象的定位。

该建筑采用中国最流行的结构系统——双筒结构，核心筒被开放空间替代，观光电梯和会议室通过悬臂伸出到了空中。挑空天井空间可以让阳光照进内廊，烟囱效应也可以加速自然通风效果。虽然这与中国典型的双筒结构系统很相似，但太平金融大厦的钢筋混凝土框架双筒结构系统仍然有其独到之处。这里的内筒没有承重墙或支撑，因

竣工时间：2014年12月
高度：228 m (748 ft)
层数：48
面积：131 281 m² (1 413 097 ft²)
主要功能：办公
业主：中国太平保险集团公司；太平人寿保险有限公司深圳分公司；太平财产保险有限公司
开发商：太平置业（深圳）有限公司
建筑设计：株式会社日建设计；深圳建筑设计研究总院有限责任公司
结构设计：株式会社日建设计；深圳建筑设计研究总院有限责任公司
机电设计：株式会社日建设计；深圳建筑设计研究总院有限责任公司
总承包商：中国建筑第三工程局有限公司

meeting spaces cantilevering into the void. The void space allows sunlight to penetrate into the inner corridors, and its chimney effect accelerates natural ventilation. While similar to the typical Chinese double-tube structural system, the reinforced-concrete bending frame double-tube system at Taiping Finance Tower is quite unique, in that the inner tube has no bearing wall or brace, making the inner void space accessible to transportation, communication spaces, natural daylight and ventilation.

At the urban scale, the Central Park situated at the intersection of the major axes of the central business district acts as an "active void" within the city. At once a traffic junction, it also serves as a vital element of the social and urban environmental infrastructure. Likewise, the building acts as an "active void," with its internal space providing for movement, ventilation, daylight and communication, connected via sunken garden and entrance hall, both to the park and to the subway and high-speed rail stations at basement level.

Shenzhen is located in the subtropics, and its warm climate allows for natural ventilation to be utilized over a long time period. Natural ventilation simulations indicated that 2 to 3 times as much natural air flow could move through the building, by the combined action of the slit openings in the outer curtain wall and the stack effect in the active void, as through operable windows on a conventionally constructed office building. It also moderates the excessive air volume during strong winds, allowing a longer period of use without the risk of windows blowing out.

Shenzhen regulations require that buildings have windows for natural ventilation. Approved by the City Planning Authority and confirmed by actual measurements post-completion, this design exceeds safety and ventilation effectiveness requirements with a consistent and sophisticated façade.

此可以通过内部的天井空间实现交通、交流、自然采光和通风等功能。

市民广场坐落在中央商务区的主要轴线的交汇处，是城区内一片"活动空间"。作为曾经的交通要道，市民广场也是重要的社会和城市环境基础设施。同样，太平金融大厦也具有它的"活动空间"，即挑空天井，内部提供各种活动、交流的场所及通风、采光的路径，通过下沉广场与入口大厅，可以从负一层直达公园与地铁和高铁站。

深圳地处亚热带，气候温暖，可长期使用自然通风。自然通风模拟试验表明，通过在外幕墙设置的通风器与挑空空间的烟囱效应，该建筑的通风量可达到自然空气流动的2至3倍，与通常的开启窗比起来，可以得到同等以上的通风量。在强风天气中，这样的设计可以减少过量空气流通，使得住户可以在较长的时间里获得通风效果，而不用担心窗户掉落。

深圳当地节能法规要求建筑物必须设置一定面积以上的自然通风的开启窗。但本设计通过稠密而精致的外墙做法和通风模拟试验，无论在安全性还是通风上都效果奇佳，论证通过了相关部门批准，并在完工后的实际测量中得到了确认。

## "The Taiping Finance Tower epitomizes contemporary ideas of a more open and transparent workplace, as well as access to natural daylight for the maximum number of occupants."

"深圳太平金融大厦最大程度地为居住者引入自然光，集中体现了营建更加开放且透明的工作场所的当代思想。"

Design Juror　设计评委

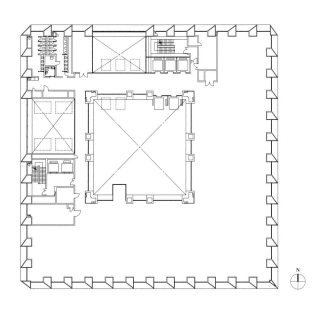

**Previous Spread | 前页**

**Left:** Overall view in context
左图：全景

**Current Spread | 本页**

**Top Left:** Detail view of façade
上左图：立面细部

**Top Right:** Interior view of entrance lobby looking up through the void space
上右图：入口大堂与中庭

**Bottom Left:** Typical floor plan
下左图：标准层平面图

# Jury Statement | 评委会评语

The design of Taiping Finance Tower suggests both movement and solidity, which is precisely the impression one should get from a major financial company building plugged so directly into the high-speed rail and local transport networks. The interplay of voids, grids, and angles enables comfortable flow of fresh air and people.

太平金融大厦的所有设计都在传递着动感和稳定的形象，这正是一个金融公司的大楼应该展现的风范。大厦直接与高速铁路及地方交通网络相连接。建筑的孔洞、栅格和各种建筑几何角度相互作用，使得新鲜空气得以流通，人流也能够畅通无阻。

# China Best Tall Building – Honorable Distinction
# 中国高层建筑荣誉奖

## Zhengzhou Greenland Plaza
## 郑州绿地中心·千玺广场

Zhengzhou | 郑州

Zhengzhou Greenland Plaza is located in the CBD core area of Zhengdong New District. It is a mixed-use development, including programs of commercial, office, hotel, and tourism services. Overall, the design of the project concentrates on resolving issues of cultural connotation and urban landmark design.

The main tower is located on a pentagonal podium, and forms a strong central axis along with the traffic island of the central square. This axis resembles a spatial sequence of traditional architecture in China, with the central square in the front, the podium in the middle, and the main tower in the background. The main tower, containing office spaces and hotel guestrooms, is designed as a perfect hexadecagon, with a core in the middle for elevator shafts and mechanical rooms. Its form is a response to the surrounding development's shape, centered on a manmade lake.

Conceived as a classical column that nevertheless is derived from scientific calculations appropriate to the

郑州绿地中心·千玺广场位于郑东新区CBD核心区内，是一个综合开发项目，包括商业、办公、酒店和旅游服务等功能。总体而言，该项目的设计关注在文化内涵和城市地标设计上。

主体建筑位于五边形裙楼之上，沿着中央广场的交通岛构筑了一条明晰的中轴线。轴线形式与中国传统建筑的空间顺序相似：中央广场在前，中间是裙楼，主体建筑位于后方。主体建筑包括办公区和酒店客房，呈正十六边形，中间有电梯和机械室。项目的造型与周边以人工湖为中心的环境相映成趣。

建筑整体造型灵感源于嵩岳寺塔，但其结构参数使用了21世纪的科学计算。最抓人眼球的是其精致的三五层楼高的轻型铝面板。它们设置了精密计算出来的向外斜面，能够增强内部光照，从而保护全玻璃外壳避免吸收过多的日光。随着楼层的上升，面板的倾斜也发生着动态变化，赋予整体建筑一种精致的质感。面板和幕墙之间留有1~2 m的空隙供清洗使用。孔隙结构让整个建筑实现了"横

**Completion Date:** July 2014
**Height:** 280 m (919 ft)
**Stories:** 60
**Area:** 240,169 sq m (2,585,158 sq ft)
**Use:** Hotel / Office
**Owner/Developer:** Greenland Group
**Architect:** Skidmore, Owings & Merrill LLP; ECADI
**Structural Engineer:** Skidmore, Owings & Merrill LLP; ECADI
**MEP Engineer:** Skidmore, Owings & Merrill LLP; ECADI
**Main Contractor:** Zhejiang Zhong Tian Construction Group Co., Ltd.
**Other Consultants:** Dunacn & Miller Design

竣工时间：2014年7月
高度：280 m (919 ft)
层数：60
面积：240 169 m² (2 584 158 ft²)
主要功能：酒店／办公
业主／开放商：绿地集团
建筑设计：SOM建筑事务所；华东建筑设计研究总院
结构设计：SOM建筑事务所；华东建筑设计研究总院
机电设计：SOM建筑事务所；华东建筑设计研究总院
总承包商：浙江中天建设集团有限公司
其他顾问方：上海达克米勒设计咨询有限公司

> "The Zhengzhou Greenland Plaza has the effect of both an ancient pagoda situated near water, and a kind of barely fathomable machine from the future."
>
> "郑州绿地中心·千玺广场坐落于河畔，既有古代宝塔的效果，又有来自未来的某种难以想象的机器的感觉。"
>
> Design Juror　设计评委

### Jury Statement | 评委会评语

The intersection of advanced science and time-tested form, as seen in the Zhengzhou Greenland Plaza, reminds us that the gap between the "intuition" that gave us the pleasing and timeless form of ancient monuments and today's "high-tech" practices is perhaps not so wide.

郑州绿地中心将先进的科技和经过时间检验的建筑形式结合起来，这令我们想起一直以来传统与技术之间存在的隔阂。给我们带来永恒美感的古老建筑形式与当今"高新技术"的运用，这两者之间的鸿沟或许并没有那么大。

**Previous Spread | 前页**

**Left:** Overall view of tower in urban context
左图: 全景

**Current Spread | 本页**

**Left:** Detail view of the façade
左图: 立面细部

**Right:** Interior view of the hotel atrium looking up
右图: 酒店室内仰视

**Opposite Right:** Section drawing showing how light reflects into the hotel atrium
对侧右图: 建筑剖面图，显示了光线是如何反射进入酒店中庭的

21st century, the tower is instantly recognizable for its sophisticated three- to five-story-tall, light-gauge painted aluminum screens. Configured at an outward cant that enhances interior daylighting through calculated reflections, the screens protect the all-glass enclosure from solar gain. The rhythmic cant of the screens, combined with their decreasing size as they rise on the building, creates a dynamic movement that gives the building a fine-grained texture. The screens are located between one and two meters from the building's curtain wall – allowing window washing to occur behind the screens. Their visual porosity varies depending on a viewer's location. When close to the building's base, the tower appears to be primarily metal; from a distance, the panels are more open and the building's glassier nature is revealed.

The entire experience of the tower is suffused with gradations of light. At the tower's entrance, the circular ground-floor office lobby features floor-to-ceiling windows, offering views of the gardens and the lake beyond. At the 40th floor, the hotel lobby anchors a 19-story atrium; natural light permeates throughout . Above the observation deck, the heliostat – a digitally shaped solar reflector – focuses sunlight into the atrium. Through an intensive series of daylighting studies, the device was specifically designed to maximize the amount of natural light entering the building. The device allows daylight to be reflected and focused into the atrium, whose surfaces are finished to help drive light deep into the space. Computer-controlled dimmer switches modulate the light level, based on the illumination provided by the reflector, enabling the atrium to consume less energy and generate less heat throughout the year. The building also features a smart-control system that utilizes external wind pressure and internal stack effects to create a well-ventilated environment.

看成岭侧成峰"的效果。站在楼底附近，整个建筑看起来以金属为主；而从远处看，建筑的玻璃幕墙就会显现出来。

　　大厦的整体感觉会随着光线的变化而变化。大厦入口处，通过一楼圆形办公大厅的落地窗，可以看到远处的花园和湖水。从酒店40层起，有一个19层高的中庭，自然光直透进来。观景台上，定日镜——一个数字成像的太阳能反射镜会将阳光聚焦到中庭。通过一系列的采光研究，该设备进行了特别制造，以使其能够最大限度地接收到自然光。设备表面设置十分精巧，可以令日光深入到空间内部，反射并聚焦到中庭。反射器可以根据照明状况自动调光，降低中庭的年度能量消耗和热量生成。该建筑还具有智能控制系统，可以利用外部的风压和内部的叠加效应，创造出良好的通风环境。

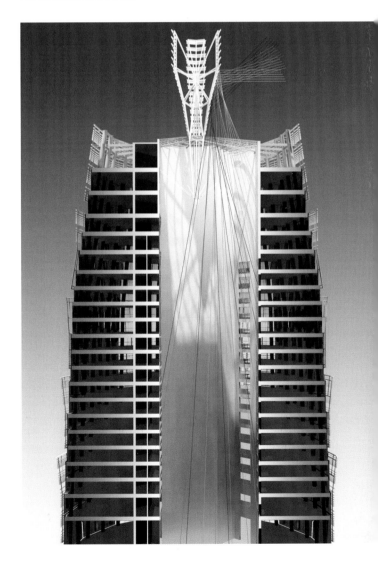

# China Best Tall Building – Nominee
中国最佳高层建筑奖提名作品

## 5 Corporate Avenue | 企业天地中心5号楼
Shanghai | 上海

5 Corporate Avenue is part of a mixed-use project that comprises the center portion of the Taipingqiao master plan. Thus, the design seeks to create a unique architectural image and retail environment, while retaining a coherence with the existing buildings in the gentrified low-rise areas of Xintiandi and preserving spatial continuity along Taipingqiao Park.

The tower was pushed to the north side of the site, while extending a three-story retail podium to the south, facing the park. This massing maintains an appropriately-scaled street wall, and reduces glare impact on the residential buildings to the south and the east. The diamond-shaped profile also enlarges the view of, and daylight to the buildings on the north, breaking down the square volume of the tower, while maintaining a high-efficiency office floor plate. The curtain wall design was inspired by traditional Chinese screens. Within this, glazed panels are shaded with aluminum fins, to create a unique vertical pattern, day and night.

**Completion Date:** January 2014
**Height:** 135 m (441 ft)
**Stories:** 27
**Area:** 67,095 sq m (722,205 sq ft)
**Use:** Office
**Owner/Developer:** Shanghai Xintiandi Management Limited
**Architect:** Kohn Pedersen Fox Associates; ECADI
**Structural Engineer:** Thornton Tomasetti; ECADI
**MEP Engineer:** Meinhardt; ECADI
**Main Contractor:** Shanghai Construction No.1 (Group) Co., Ltd.
**Other Consultants:** Arup (LEED); Design Land Collaborative (landscape); Meinhardt façade Technology (Shanghai) Ltd (façade); RTKL Ltd (interiors); Shanghai Research Institute of Building Sciences (sustainability)

上海企业天地中心5号楼是以太平桥街区建设计划为主体的综合项目的一部分。因此，设计在追求创造独特建筑形象和良好商业环境的同时，还要与新天地现有更新的低层建筑保持协调，以维持太平桥街区的空间连续性。

大楼位于基地北端，它向南延伸出了一座面朝公园的三层商铺裙楼。项目包括建造一道高度适中且面向街道的墙体，以减弱东部与南部居民区的强光照晒。菱形的外观不但扩展了楼北面的视野，改善了采光，也改变了大楼的方形设计，同时保持了办公区域的使用效率。幕墙的设计灵感来源于中国传统的屏风。幕墙外部设置的一系列不同角度的外伸装饰条，与竖向的铝板窗间墙一起，共同起到遮挡直射阳光和减少幕墙反射光污染的效果，无论白天还是夜晚都创造出别具一格的竖向单元效果。

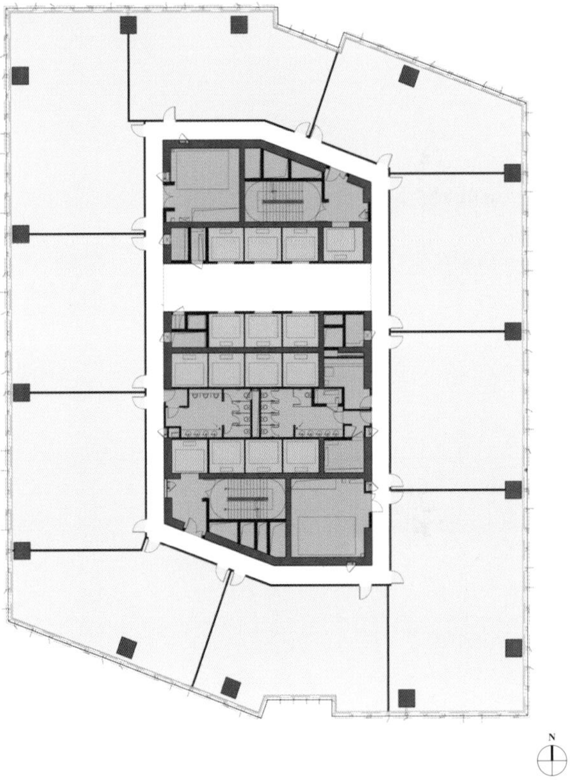

竣工时间：2014年1月
高度：135 m（441 ft）
层数：27
面积：67 095 m²（722 205 ft²）
主要功能：商业/办公
业主/开发商：上海新天地商业管理有限公司
建筑设计：KPF建筑事务所；华东建筑设计研究总院
结构设计：宋腾添玛沙帝工程顾问公司；华东建筑设计研究总院
机电设计：迈进工程设计咨询有限公司；华东建筑设计研究总院
总承包商：上海建工一建集团有限公司
其他顾问方：奥雅纳工程咨询有限公司（可持续性）；地茂景观设计咨询（上海）有限公司（景观）；迈进外墙建筑设计咨询（上海）有限公司（外墙）；RTKL建筑事务所（室内设计）；上海建筑科学研究院（绿色建筑）

**Opposite Left:** Overall view of tower in context
对侧左图：全景

**Top Left:** Detail view of the façade's glazing and aluminium fins
上左图：立面幕墙的细部

**Top Right:** Interior view of the entrance lobby
上右图：入口大堂

**Bottom Right:** Typical floor plan
下右图：标准层平面图

# China Best Tall Building – Nominee
## 中国最佳高层建筑奖提名作品

## Agile Center | 雅居乐中心
Guangzhou | 广州

The Agile Center is a single-volume high-rise building, without a podium, but part of a larger master-planned development. Standing shoulder to shoulder with the Pearl River City to the east, its main body, in the distinctive shape of an oval, is oriented southwest towards a major traffic circle, forming the main entrance to the plaza. The land to the north is vacated to form the entrance to a bank and the traffic circle. Besides, a north-south pedestrian overpass connects the second floor to neighboring plots and serves as a marker for the building, creating a three-plaza system, including the B1-5 Project to the north and the F1-1 Project to the south.

A series of façade views are created by the oval geometric structural form, as if it were embracing the colorful city life with open arms. This also dictates the shape of the entrance plaza. The pedestrian overpass, arranged along the garden landscape in an arc, echoes the nearby Pearl River's flow along the southern edge of the development, not only intensifying the continuous experience for pedestrians, but also creating a visual focus for the project's corner, as well as energizing the northeast sector. The oval shape of the building provides a unique and changing experience for pedestrians and vehicles. Through the use of various construction materials, including metal, stone, glass and horizontal and vertical sunshade devices, the building is environmentally friendly and energy saving. The external façade presents various appearances at different times of day, in different seasons, and from various viewing angles. After a rigorous review of options, a gracious and classic modern building, bearing solid and generous characteristics, has been brought into the new development.

雅居乐中心是单栋超高层塔楼。建筑不设裙楼，主体呈椭圆形，朝向西南，面对并退让道路交叉口，形成主入口广场，并与东侧的珠江城超高层建筑比肩而立。在用地的北面空出用地作为银行入口广场和交通集散广场。二层设置南北向的步行天桥连接邻近地块，并局部放大，和北侧的 B1-5 地块项目，以及南边的 F1-1 地块项目连通，形成一个立体的广场体系。

椭圆形的几何建筑形式创造了一个连续的沿街面，如同双臂展开拥抱都市精彩的生活，同时平面布置方位形成入口广场。顺着弧线园林景观布置的天桥，呼应着该发展地区南方的珠江流线，创造出立体的、流畅的自然动线，强化了天桥连续性的人行体验，并创造出项目的视觉焦点，带动地块东北侧的活力。椭圆形的塔楼形态，提供了独特并瞬息万变的人行和车行体验，通过金属、石材、玻璃和横竖遮阳板装置等不同建筑材料的运用，这一建筑因而具备了环保和节能性能，建筑外立面从而随着不同的时间、不同的日期，甚至不同的观看角度，呈现出多样化的容貌。综观建筑各构成单元的组合，为珠江新城都市发展带来的是一座精致典雅，又不失稳重大方的现代化建筑。

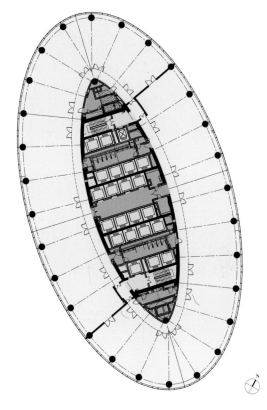

**Completion Date:** December 2014
**Height:** 190 m (622 ft)
**Stories:** 39
**Area:** 115,932 sq m (1,247,882 sq ft)
**Use:** Office
**Owner/Developer:** Agile Group
**Architect:** Skidmore, Owings & Merrill LLP
**Structural Engineer:** Skidmore, Owings & Merrill LLP
**MEP Engineer:** Guangzhou Design Institute
**Main Contractor:** Agile Group

竣工时间：2014 年 12 月
高度：190 m (622 ft)
层数：39
面积：115 932 m² (1 247 882 ft²)
主要功能：办公
业主 / 开发商：雅居乐集团
建筑设计：SOM 建筑事务所
结构设计：SOM 建筑事务所
机电设计：广州市设计院
总承包商：雅居乐集团

**Opposite Left:** Overall view of building
对侧左图：全景
**Top Left:** Pedestrian overpass from below
上左图：天桥仰视
**Top Right:** Detail view of the façade
上右图：建筑立面细部
**Bottom Right:** Typical floor plan
下右图：标准层平面图

# China Best Tall Building – Nominee
中国最佳高层建筑奖提名作品

## Changsha Xinhe North Star Delta Center
长沙北辰新河三角洲

Changsha | 长沙

The combination of the urban context and historical origin drove the urban design concept of the "Landscape Island City," resulting in a multi-layered shaping strategy that provides a rich urban landscape and maximizes its riverside location with a cascading green platform. The uniquely shaped office tower is the focus of the whole complex, which includes a compatibly-shaped mid-rise hotel tower, and forms the backdrop of Xinhe Delta Park. The multi-colored, multi-textured glass of its exterior represents a special geo-morphological feature of floodplains.

The triangular plan and rounded edges of the office tower reduce the surface area of the building and thus its wind resistance, thereby significantly reducing the cost of its structure. A unitized curtain-wall system increases the efficiency and quality of its construction. The multi-layered curtain wall emphasizes shading on the south side and high-performance glass on the east and west sides, reducing heat absorption and increasing comfort, while symbolizing an abstracted, cascading rock physiognomy.

**Completion Date:** September 2014
**Height:** Office: 235 m (771 ft); Hotel: 117 m (383 ft)
**Stories:** Office: 46; Hotel: 26; Retail: 7
**Area:** 314,337 sq m (3,383,495 sq ft)
**Use:** Office / Hotel / Retail
**Owner/Developer:** North Star Real Estate Ltd.
**Architect:** RTKL; Jerde Partnership; Beijing Institute of Architectural Design
**Structural Engineer:** Beijing Institute of Architectural Design
**MEP Engineer:** Beijing Institute of Architectural Design
**Main Contractor:** Beijing Construction Engineering Group
**Other Consultants:** ACLA (landscape); Jerde Partnership (planning); Shanghai Kighton Façade Consultants Co., Ltd. (façade)

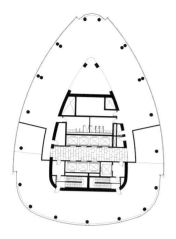
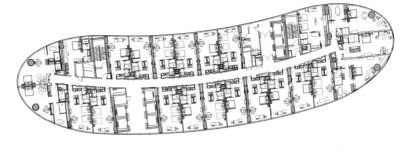

项目结合了城市文脉与历史源起，催生出长沙"山水洲城"的设计理念，以多层次的造型手法，创造出丰富灵动的城市景观，最大限度地引入河岸景观和层叠的绿化平台景观。其中设计标新立异的办公塔楼是整个建筑群的核心，连同设计新颖和谐的中高层酒店建筑共同构成了新河三角洲这一城市新区。其外表多彩多样的玻璃是河漫滩特殊地形的展示。

办公塔楼的三角设计和圆形边缘减少了建筑表面积，也因此减轻了风阻力，显著降低了结构的成本。整体幕墙系统也提高了建筑效率和质量。多层幕墙提高了南墙的遮阴能力，东面和西面的高性能玻璃减少了热量吸收，增强了舒适性。总体来说，建筑造型象征着一个分离出来的层叠状岩石地貌。

**Opposite Left:** Overall view of towers in urban context
对侧左图：塔楼全景

**Top Left:** Typical office floor plan
上左图：办公标准层平面图

**Top Right:** Typical hotel floor plan
上右图：酒店标准层平面图

**Bottom:** Hotel entry
下图：酒店入口

竣工时间：2014 年 9 月
高度：办公楼：235 m (771 ft)；酒店：117 m (383 ft)
层数：办公楼：46；酒店：26；商业：7
总面积：314 337 m² (3 383 495 ft²)
主要功能：办公 / 酒店 / 商业
业主 / 开发商：北辰房地产有限公司
建筑设计：RTKL 建筑事务所；捷得国际建筑师事务所；北京市建筑设计研究院
结构设计：北京市建筑设计研究院
机电设计：北京市建筑设计研究院
总承包商：北京建工集团
其他顾问方：傲林国际设计有限公司（景观）；捷得国际建筑师事务所（规划）；上海凯腾幕墙设计咨询有限公司（幕墙）

# China Best Tall Building – Nominee
中国最佳高层建筑奖提名作品

## Changzhou Modern Media Center
常州现代传媒中心

Changzhou | 常州

Changzhou Modern Media Center is located in the Xinbei district of Changzhou, which is a newly developed urban area. The form of the tower is inspired by the 1,300-year-old Tianning Temple, which is the most famous historic building in Changzhou. The project intends to represent a pagoda, which in traditional Chinese culture is thought to bring good fortune. The lower part of the main tower is office space, and the higher section is a Marriott hotel. The top of the tower consists of luxury apartments and an observatory.

The complex also contains other functions, including radio and TV broadcasting, retail, and a theater. To avoid the negative impact of a superblock on the surrounding area, the tower's podium is divided into smaller blocks, forming internal streets that can be navigated at the pedestrian scale. The placement of the tower on the site is intended to maximize its contribution to the urban landscape, and to reduce the blockage of light to the residential buildings beyond.

**Completion Date:** 2013
**Height:** 256 m (840 ft)
**Stories:** 58
**Area:** 89,767 sq m (966,244 sq ft)
**Use:** Residential/Hotel/Office
**Owner:** Changzhou Broadcasting Station
**Developer:** Changzhou Radio and TV Realty Company, Ltd.
**Architect:** Institute of Shanghai Architectural Design & Research, Co., Ltd.
**Structural Engineer:** Institute of Shanghai Architectural Design & Research, Co., Ltd.
**MEP Engineer:** Institute of Shanghai Architectural Design & Research, Co., Ltd.
**Main Contractor:** China Construction Third Engineering Bureau Co., Ltd.

常州现代传媒中心位于常州市的新开发区——新北区。这座塔楼的设计灵感来源于常州最著名的古建筑——拥有1 300年历史的天宁寺。该建筑被设计成了宝塔形，在中国传统文化中，宝塔被认为可以带来好运和福气。建筑的低层部分是办公区域，高层部分是万豪酒店，顶部则建有豪华套房和一个观景台。

这座综合建筑还拥有其他功能，包括电台、电视台、商场以及电影院。为了避免占地过大给周边环境带来负面影响，该建筑的裙楼被设计分割为了多个小型体块，从而形成了类似步行街的内部街道。项目的总体布局旨在尽可能地展现其城市景观效果，同时尽量避免遮挡后方居民楼的阳光。

竣工时间：2013 年
高度：256 m（840 ft）
层数：58
面积：89 767 m²（966 244 ft²）
主要功能：住宅 / 酒店 / 办公
业主：常州广播电台
开发商：常州广播电视实业有限公司
建筑设计：上海建筑设计研究院有限公司
结构设计：上海建筑设计研究院有限公司
机电设计：上海建筑设计研究院有限公司
总承包商：中国建筑第三工程局有限公司

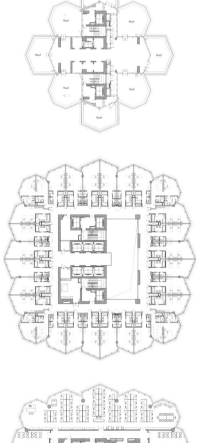

**Opposite Left:** Overall view from south
对侧左图：南部全景
**Top Left:** Interior view of a TV editing office
上左图：电视制作中心内景
**Top Right:** Observatory floor plan
上右图：天文台平面图
**Middle:** Typical hotel floor plan
中图：酒店标准层平面图
**Bottom:** Typical office floor plan
下图：办公标准层平面图

# China Best Tall Building – Nominee
中国最佳高层建筑奖提名作品

## Chongqing Land Group Headquarters
重庆地产集团总部

Chongqing | 重庆

Located in Chongqing, this building is inspired by the city's surrounding mountain context. As a result, the design takes on a monolithic quality that appears steady and unchanging. Massive in appearance, the mega-structure is divided into two parts. The southern section houses the Chongqing Land Group headquarters, while the northern part is rentable office space. Additionally, conference, catering, and fitness facilities are located within the development.

A large aperture divides the two sections of the building, creating a ring shape. This design provides a variety of public spaces, including different levels of green platforms, which create social spaces for office users and the general public. Many of these outdoor spaces are integrated into the structure, offering natural shade. The building's façade is defined by a series of narrow vertical mullions, which both augment the verticality of the structure and root it firmly into the ground.

**Completion Date:** March 2014
**Height:** 99 m (323 ft)
**Stories:** 20
**Area:** 108,442 sq m (1,167,260 sq ft)
**Use:** Office
**Owner:** Kangtian Real Estate Co., Ltd; Chongqing Land Group
**Developer:** Kangtian Real Estate Co., Ltd.
**Architect:** Beijing Institute of Architectural Design (design)
**Structural Engineer:** Beijing Institute of Architectural Design (design)
**MEP Engineer:** Beijing Institute of Architectural Design (design)
**Main Contractor:** No.3 Construction Co., Ltd. of Chongqing Construction Engineering Group

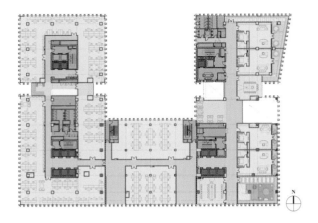

这栋坐落于重庆的建筑其设计灵感来自周边的山川风貌，因此，整个建筑体现出山峦一般的稳固与恒久。这个庞然大物的外观宏伟壮丽，分为南北两部分。南部为重庆地产集团总部，北部则是写字楼，其间还包括会议室、餐饮服务以及健身场所等功能设施。

一道巨大的缝隙将这栋环形建筑一分为二。项目规划了多种多样的公共空间，包括错落有致的绿化地带，这为办公人员和公众提供了宽阔的社交空间。很多户外空间与大楼整体结构相融合，形成了天然的荫蔽。大楼的外立面由一系列狭长的垂直框架组成，既增强了巍峨耸立的垂直感，又能使其深深扎根于大地。

竣工时间：2014 年 3 月
高度：99 m（323 ft）
层数：20
面积：108 442 m²（1 167 260 ft²）
主要功能：办公
业主：康田置业有限公司；重庆地产集团
开发商：康田置业有限公司
建筑设计：北京市建筑设计研究院
结构设计：北京市建筑设计研究院
机电设计：北京市建筑设计研究院
总承包商：重庆建工集团第三建筑有限公司

**Opposite Top:** Overall view in context
对侧上图：全景
**Opposite Bottom:** Building entry
对侧下图：建筑入口
**Top Left:** Floor plan – level 17
上左图：17层平面图
**Right:** Interior common space
右图：室内公共空间

# China Best Tall Building – Nominee
## 中国最佳高层建筑奖提名作品

# Forebase Financial Plaza | 申基金融广场
Chongqing | 重庆

Forebase Financial Plaza is located in the heart of Chongqing's Yuzhong Peninsula, the city's traditional business center. The scheme consists of three interconnected masses and a podium that houses office space, a hotel, and a park, among other facilities. The reinforced concrete frame and core-wall constitute the main structural system of the building. The design uses vertical lines on the glass curtain wall, embodying the tall and straight architectural aesthetic of tall buildings. A series of dark-grey metal mullions are implemented to further achieve this effect.

The consistent motif of vertical lines on the facade is broken by a five-meter cantilevered glass swimming pool and sky garden located 230 meters above the ground. The hotel lobby is set on the 51st floor, which has a great view of the river and mountains. Platforms, steps, and water features, as well as tall trees and native plants, are all incorporated into the landscape to reflect the surrounding mountainous context.

**Completion Date:** April 2014
**Height:** 245 m (805 ft)
**Stories:** 54
**Area:** 205,915sq m (2,216,527 sq ft)
**Primary Functions:** Hotel / Office
**Owner/Developer:** Forebase Group
**Architects:** Tanghua Architect & Associates Co., Ltd.; Chongqing Architecture and Design Institute
**Structural Engineer:** Chongqing Architecture and Design Institute
**MEP Engineer:** Chongqing Architecture and Design Institute
**Main Contractor:** MCC Huaye Resources Development Co., Ltd.
**Other Consultants:** Elmich (Guangzhou) Landscape Engineering Co., Ltd. (green walls); Lutron (lighting); METROSTUDIO (landscape); Shenzhen Catic Curtain Wall Engineering Co., Ltd. (façade)

申基金融广场坐落于重庆传统商业中心——渝中区的核心地段。项目包括一高两矮三个塔楼和裙房，容纳了办公、五星级酒店、名品商业和相关配套设施。建筑主体结构为型钢混凝土框架—钢筋混凝土筒体结构。造型上采用竖向线条的玻璃幕墙，充分体现超高层建筑高耸挺拔的建筑美感。深灰色的金属竖直挺拔，强化了这种视觉效果。

在建筑顶部出挑的无边际泳池和距地面 230 m 的空中花园，打破了全竖向线条的单一感。酒店大堂设置于建筑的 51 层，可以饱览重庆的江景山色。建筑内部商业街、人行步道和建筑周围的台地、阶梯、广场构成一体，充分体现了重庆的山城风貌。

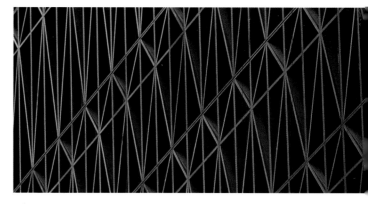

竣工时间：2014 年 4 月
高度：245 m（805 ft）
层数：54
面积：205 915 m²（2 216 527 ft²）
主要功能：酒店 / 办公
业主 / 开发商：申基集团
建筑设计：汤桦建筑设计事务所；重庆市设计院
结构设计：重庆市设计院
机电设计：重庆市设计院
总承包商：中冶集团华冶资源开发有限责任公司
其他顾问方：Elmich（广州）景观设计有限公司（垂直绿化设计）；路创电子公司（照明）；意大利迈丘设计（景观）；深圳中航幕墙工程有限公司（幕墙）

**Opposite Left:** Overall view from south
对侧左图：南向全景
**Top Left:** "Sky pool"
上左图：空中泳池
**Top Right:** Typical floor plan
上右图：标准层平面图
**Bottom:** Detail view of the façade
下图：立面细部

# China Best Tall Building – Nominee
## 中国最佳高层建筑奖提名作品

## Fortune Financial Center | 北京财富金融中心
Beijing | 北京

Located along Beijing's Third Ring Road opposite CCTV Headquarters, Fortune Financial Plaza completes a development that already features offices, apartments, a 5-star hotel and retail. Beijing's second-tallest building completion in 2014, this office tower is set in a large and extensively landscaped park, linked to the city's Metro network via subterranean retail arcade. The tower's form acknowledges and echoes the design of the buildings in the area, with inclined façade planes and curved corners; yet it is scaled, articulated and refined to reflect its importance. Clad with silver-colored aluminum and reflective glass fins in double-glazed units, a taut and elegant skin defines the carefully sculpted tower.

A main entrance lobby, finished in imported granite and marble, is enclosed by a clear glass screen, linking it visually and physically to the extensive park outside. Enriched with water walls and artwork, it provides an appropriately scaled and elegant foyer to the tall tower. Column-free office space, with a generous floor-to-ceiling height, provides exclusive accommodation.

**Completion Date:** May 2014
**Height:** 267 m (876 ft)
**Stories:** 60
**Area:** 151,585 sq m (1,631,647 sq ft)
**Use:** Office
**Owner/Developer:** Xiang Jiang Xing Li Estates Development Ltd.
**Architect:** P & T Group; CERI, Ltd.
**Structural Engineer:** Arup; CERI, Ltd.
**MEP Engineer:** WSP | Parsons Brinckerhoff; CERI, Ltd.
**Other Consultants:** ACLA (landscape); Environmental Market Solutions, Inc. (LEED); Schmidlin (façade); WT Partnership (cost)

财富金融中心坐落于北京三环路一侧,正对CCTV总部大楼,现在已发展为集办公、住宅、五星级酒店和商铺等功能于一体的综合性建筑。这座办公楼是北京市第二高的建筑,于2014年竣工,位于地域广阔的公园景观之中,通过地下商铺长廊与城市地铁网络相连。

这座高楼的外形继承延续了该区域现有大楼的设计特色,具有倾斜的外立面和圆角结构,但其外形经过了调整、组织与美化,从而彰显出其重要地位。双层玻璃窗上安装了银色铝合金和反光玻璃片,让这座精细雕琢的建筑拥有了紧致优雅的外皮。

主入口的大堂以进口花岗岩和大理石饰面,四周被一片透明的玻璃屏围绕着,从这里不仅能看到外面的公园,还能直接走入其中。水墙和艺术品丰富了前厅的内涵,为这座高层建筑提供了一个大小适中的优雅休息场所。楼内的无柱办公室层高很大,为租户提供了独一无二的体验。

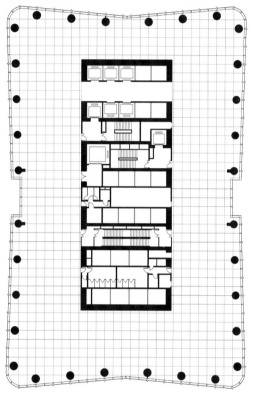

竣工时间:2014年5月
高度:267 m (876 ft)
层数:60
面积:151 585 m² (1 631 647 ft²)
主要功能:办公
业主/开发商:香江兴利房地产开发有限公司
建筑设计:巴马丹拿建筑设计咨询有限公司;中冶京诚工程技术有限公司
结构设计:英国奥雅纳工程顾问公司(设计);中冶京诚工程技术有限公司(记录工程师)
机电设计:科进|柏诚集团;中冶京诚工程技术有限公司
项目管理:香江国际中国地产有限公司
其他顾问方:傲林国际设计有限公司(景观);EMSI(LEED);Schmidlin公司(外立面);务腾咨询有限公司(工程造价)

**Opposite Left:** Overall view from east
对侧左图:东立面全景

**Top Left:** Building entry
上左图:建筑入口

**Top Right:** Overall view in context
上右图:全景

**Bottom:** Ground level floor plan
下图:一层平面图

# China Best Tall Building – Nominee
## 中国最佳高层建筑奖提名作品

## HVW Headquarters | 台湾HVW总部
Taoyuan | 桃园

The HVW Headquarters' design is informed by its location within an industrial neighborhood adjacent to a freeway interchange. Its proximity to the highway gives this building the opportunity to act as a billboard and gateway for the city, using world-class architecture to advertise its presence to daily commuters. The building presents an assertive and distinctive physical form, with abstract volumes floating within an industrial landscape.

Office blocks of various sizes and orientations lie along the core axis of the building. These boxes have large openings of dark-colored glass inset into concrete frames. Openings are carefully arranged to make the best use of sunlight. Generous north–south windows allow natural light into interior office spaces, while east–west exposures are limited by solid concrete walls. The extensive use of concrete cladding recalls Brutalist typologies and reflects the industrial nature of the building's context, while the distinctive block patterning is meant to call to mind the form of the shipping containers that are so prevalent in the area.

**Completion Date:** January 2014
**Height:** 83 m (271 ft)
**Stories:** 19
**Area:** 54,916 sq m (591,111 sq ft)
**Use:** Office
**Owner/Developer:** Kuang Chun Cheng Construction
**Architect:** EHS ArchiLab + Hsuyuan Kuo Architects & Associates; Lu Wen Cheng Architects & Associates
**Structural Engineer:** Tai Yun Fa Structural Consultants
**MEP Engineer:** Cheng Yuan MEP Consultants
**Main Contractor:** Chun Cheng Construction

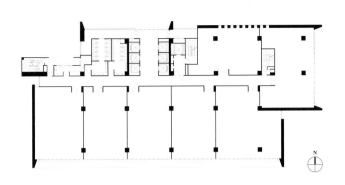

HVW 总部的灵感来自其地理位置,它位于毗邻高速立交桥的工业区内。临近高速公路的地理位置使它有机会成为城市的门户与广告牌,向每日往返的上班族展示这个世界一流的建筑设计。它自信而与众不同,漂浮着的几何建筑体块都与工业区景观相得益彰。

楼内有多个不同面积与朝向的办公区域,沿建筑中轴线排列。这些办公室装有嵌入混凝土框架的巨大深色玻璃窗。这些窗经过精心设计,充分利用了阳光。宏大的南北向窗户使自然光能进入办公区域内部,东西向的采光则被混凝土外墙限制。大量使用混凝土覆层令人联想到野兽派的作品,也反映出了所在环境的工业特质。独特的箱形样式让人不禁联想起该地区常见的船运集装箱。

竣工时间:2014 年 1 月
高度:83 m(271 ft)
层数:19
总面积:54 916 m² (591 111 ft²)
主要功能:办公
业主 / 开发商:广春成建筑公司
建筑设计:大尺设计 + 郭旭原建筑师事务所;Lu Wen Cheng 建筑师事务所
结构设计:Tai Yun Fa 结构咨询公司
机电设计:Cheng Yuan 机电咨询公司
总承包商:春成建筑公司

**Opposite:** Overall view from west
对侧图:西立面全景

**Top Left:** Typical floor plan
上左图:标准层平面图

**Top right:** 1F floor plan
上右图:一层平面图

**Right:** Detail view of box structure
右图:箱形结构细部

# China Best Tall Building – Nominee
中国最佳高层建筑奖提名作品

## J57 SkyTown | J57天空之城
Changsha | 长沙

J57 SkyTown is a skyscraper constructed by using a proprietary prefabrication system, achieving several speed milestones, such as "three stories in one day" and "57 stories in 19 days." The all-in-one manufacturing approach to building construction has yielded other impressive statistics. The building's filtration systems and 20 cm thick thermally insulated walls eliminate 99.9% of PM 2.5 particles, a major concern in China, while offering 80% more energy efficiency than similar systems in a comparable conventionally constructed building.

The essential principles of the manufacturer/designer are that its buildings be 90% factory-assembled, offer five times the energy efficiency of a typical building, and have the ability to withstand a 9.0-magnitude earthquake. The prototype outside Changsha can hold 4,600 office workers and accommodate 800 families. It is the latest in a series of thought-provoking, rapidly-assembled and highly efficient buildings on this factory campus. Its potential is cast in idealistic terms, but from the standpoint of construction efficiency, it is already a substantial achievement.

**Completion Date:** February 2015
**Height:** 208 m (682 ft)
**Stories:** 57
**Area:** 179,600 sq m (1,933,198 sq ft)
**Use:** Residential / Office
**Owner/Developer:** Sky City Investment Co., Ltd.
**Architect:** Broad Sustainable Building Co., Ltd
**Structural Engineer:** Sky City Investment Co., Ltd.; RBS Architectural Engineering Design Associates
**MEP Engineer:** Sky City Investment Co., Ltd.
**Main Contractor:** Sky City Investment Co., Ltd.
**Other Consultants:** Central Southern Geotechnical Design Institute Co., Ltd. (geotechnical); Great Earth Architects & Engineers International (landscape); National Engineering Research Center for Fire Protection (fire)

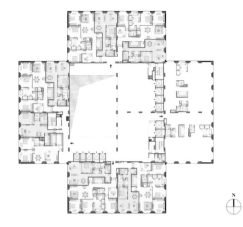

J57天空之城是一座非同一般的摩天大楼。施工过程采用了独有的预制拼装建造工艺，因此在建造速度上取得了众多里程碑式的突破，例如"1天完成3层楼"、"19天完成57层楼"这样的奇迹。这种建造方式还产生了其他令人印象深刻的数字。空气质量在中国是一个大问题，大楼的过滤系统和厚达20 cm的隔热墙可以过滤掉99.9%的PM 2.5颗粒。相比传统建筑中的同类系统，天空之城的过滤系统的能效高达80%。

建筑商和设计师们对天空之城项目的基本要求是：工程的90%在工厂组装完成，能效是常规建筑的5倍，同时能抵御9.0级强震。天空之城这一长沙市郊的原型建筑可容纳4 600名上班族工作，800个家庭生活。该建筑工厂曾参与建造了一系列令人大开眼界的建筑，一直以快速装配和高能效知名，天空之城正是其最新的工程成果。如果说对天空之城未来潜力的描述还基本停留在理想层面上的话，那么从施工效率来看，它已经是一项壮举了。

竣工时间：2015年2月
高度：208 m（682 ft）
层数：57
面积：179 60 m² （1 933 198 ft²）
主要功能：住宅/办公
业主/开发商：天空城市投资股份有限公司
建筑设计：远大可建科技股份有限公司
结构设计：天空城市投资股份有限公司；广州容柏生建筑结构设计事务所
机电设计：天空城市投资股份有限公司
工程管理：天空城市投资股份有限公司；中国建筑第五工程局股份有限公司
总承包商：天空城市投资股份有限公司
其他顾问方：中南勘察设计院股份有限公司（岩土工程）；大地建筑事务所（景观）；国家消防工程技术研究中心（消防）

**Opposite Left:** Overall view
对侧左图：全景
**Top Left:** Overall view in context
上左图：全景
**Top Right:** Typical floor plan
上右图：标准层平面图
**Bottom:** Interior view of multifunctional hall
下图：多功能厅内景

# China Best Tall Building – Nominee
中国最佳高层建筑奖提名作品

## Ji'nan Greenland Center | 济南绿地中心
Ji'nan | 济南

The site for Ji'nan Greenland Tower had several significant constraints. A telecommunications office building and buried fiber-optic cable had to be preserved, and local protected sites such as Spouting Spring, Five Dragon Pond and Da Ming Lake are nearby. Planners required 10,000 square meters of common green space. The design not only dealt with the restrictive conditions, but also took full advantage of the surrounding landscape. As Jinan's tallest building, it has become the landmark of the "City of Springs."

The design response is a beveled triangular building with large setbacks, which forms a coherent building image, composed by the ribbon-like top and vertical wings, a metaphor for the spouting springs of Ji'nan. The exterior's glass-aluminum panel curtain walls with perforated aluminum wings act as vertical solar protection devices, which reduce the heat gain of the building and prevent solar glare, as well as provide the unique facade treatment that terminates in the arched crown and supports dynamic LED lighting effects at night.

**Completion Date:** December 2014
**Height:** 303 m (994 ft)
**Stories:** 60
**Area:** 197,140 sq m (2,121,997 sq ft)
**Use:** Office
**Developer:** Greenland Group
**Architect:** ECADI
**Structural Engineer:** ECADI
**MEP Engineer:** ECADI
**Main Contractor:** Shanghai Construction Group

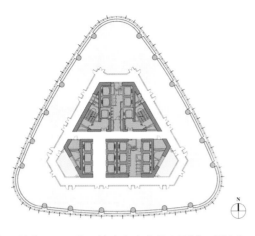

济南绿地中心项目位于济南市商业核心区域，周边有趵突泉、五龙潭、大明湖等历史文化古迹。规划要求在基地东侧建设约 10 000 m² 公共绿化，整个设计在解决了复杂限制条件的前提下，利用基地西南角留下的一块较完整的三角形用地，很自然地将超高层塔楼设计成弧线三角形，正好嵌入基地西南侧，旋转角度也与城市道路边线吻合，完整展现了超高层塔楼直接落地的挺拔形象。商业裙房沿北侧的普利街与南侧的共青团路展开，通过不同层次的连廊和屋顶的组合连接，形成几组既分又合的组群。

建筑外墙为铝合金玻璃幕墙体系，布置穿孔铝合金翼板作为竖向遮阳装置，降低了大楼的热增益并限制了太阳眩光。随着大楼向上收缩升高，这些遮阳板汇合编织在一起，在建筑顶部形成独特的开放式塔冠。塔冠飘带及塔身竖向翼一气呵成的建筑造型，也隐喻了济南泉城中涌泉特色形态。

竣工时间：2014 年 12 月
高度：303 m (994 ft)
层数：60
总面积：197 140 m² (2 121 997 ft²)
主要功能：办公
开发商：绿地集团
建筑设计：华东建筑设计研究总院
结构设计：华东建筑设计研究总院
机电设计：华东建筑设计研究总院
总承包商：上海建工集团

**Opposite Left:** Overview of tower in context
对侧左图：全景
**Top Left:** Typical office floor plan
上左图：办公标准层平面图
**Top Right:** Details of the façade
上右图：幕墙细部
**Bottom:** Interior view of the entrance lobby
下图：入口大厅内景

# China Best Tall Building – Nominee
## 中国最佳高层建筑奖提名作品

### Jing Mian Xin Cheng Tower | 京棉新城大厦
Beijing | 北京

The Jinan Greenland Center is located in the central business district (CBD) of Jinan, surrounded by historical and cultural monuments such as the Spouting Spring, Wulong Tan and Daming Lake. Its planning permission requires 10,000 square meters of common green space on the east side of the site. The design comprehensively dealt with the complex, restrictive conditions and matched the triangular area remaining on the southwest corner of the site. The tower is designed to be spherical triangle whose direction is oriented to the southwest, and whose rotation angle also matches with that of the surrounding streets. The tower occupies the center with a tall and straight image. The commercial podiums lie, wing-like, along Puli Street on the north and Gongqingtuan Road on the south. The podiums are independent volumes, but linked by layers of corridors and roofs.

The exterior of the building consists of glass-aluminum panel curtain walls with perforated aluminum wings as vertical solar protection devices, which reduce the heat gain of the building and prevent solar glare. As the tower narrows along with its altitude, the solar protection panels weave together and become a distinct crown on the top. The coherent building image composed by the ribbon-like top and vertical wings forms a metaphor for the sprouting springs in Jinan.

京棉新城的设计灵感来自于曾经位于北京四环路边上的纺织品市场（即新城选址）。大厦的织线状外立面十分引人注目，反映出了当地的历史。铝板穿孔而成的褶皱形成了特有的厚重织纹外墙，倾斜的玻璃则像"线"一样慢慢地被编进大厦。褶皱的外立面玻璃板在办公空间内侧形成了一系列窗户。

景观设计力求与周围环境融为一体，同时建筑给行人以及四环路上从远处看到这座建筑的司机与乘客带来强烈的视觉冲击。公众广场景观中的褶皱模仿了建筑外立面的设计，既为人们提供了座位，又划定了公共活动区域。同时，附近的多层建筑和购物大楼也延续了"编织"主题。

**Completion Date:** September 2013
**Height:** 102 m (335 ft)
**Stories:** 24
**Area:** 87,800 sq m (945,071 sq ft)
**Use:** Office
**Owner/Developer:** Sino-Ocean Land
**Architect:** Spark Architects
**Structural Engineer:** Beijing Institute of Architectural Design
**MEP Engineer:** Beijing Institute of Architectural Design (engineer of record)
**Main Contractor:** Yuanda
**Other Consultants:** Schmidlin (façade); Yuanda (façade)

竣工时间：2013 年 9 月
高度：102 m (335 ft)
层数：24
面积：87 800 m² （945 071 ft²）
主要功能：办公
业主／开发商：远洋地产
建筑设计：思邦建筑
结构设计：北京市建筑设计研究院
机电设计：北京市建筑设计研究院
总承包商：远大集团
其他顾问方：Schmidlin（外立面）；远大（外立面）

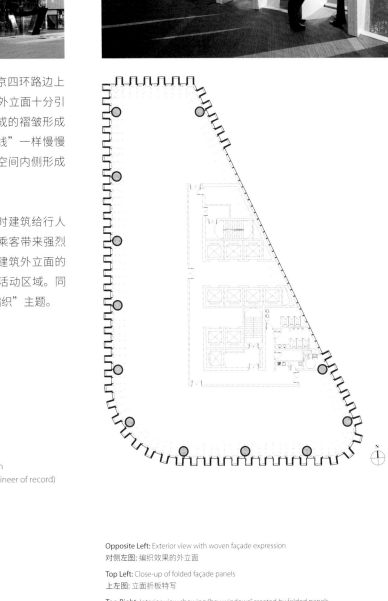

**Opposite Left:** Exterior view with woven façade expression
对侧左图：编织效果的外立面

**Top Left:** Close-up of folded façade panels
上左图：立面折板特写

**Top Right:** Interior view showing "bay windows" created by folded panels
上右图：内景，可以看出折板构成的凸窗

**Bottom:** Ground floor plan
下图：地面层平面图

# China Best Tall Building – Nominee
## 中国最佳高层建筑奖提名作品

## Kingtown International Center
### 南京金奥国际中心
Nanjing | 南京

The Hexi District of Nanjing is in the midst of transforming from a rural area to a vibrant business hub and civic center. This building is an anchor and gateway to the new district, serving as an icon of the area's new urbanity. This next-generation commercial and hotel tower maximizes performance, efficiency, and occupant experience. Its distinctive external bracing system creates more efficient lateral support – requiring less steel and an overall 20% reduction in building material.

The design of the tower is rooted in the notion of developing this parcel and the neighboring parcel to serve both as a gateway to, and as a symbol of the vitality of the new district. Its faceted form is derived from the juxtaposition of the innovative double-skin façade that creates solar shading and an insulating chamber between the building envelope and the occupied space. Vented openings in the outermost curtain wall allow wind pressure to draw built-up heat out of the cavity, lowering temperatures along the inner exterior wall.

**Completion Date:** September 2014
**Height:** 231 m (759 ft)
**Stories:** 56
**Area:** 230,621 sq m (2,482,384 sq ft)
**Use:** Hotel / Office
**Owner/Developer:** JiangSu Goldenland Real Estate Development Co., Ltd.
**Architect:** Skidmore, Owings & Merrill LLP
**Structural Engineer:** Skidmore, Owings & Merrill LLP
**MEP Engineer:** WSP | Parsons Brinckerhoff
**Main Contractor:** Wuhan Construction Engineering Group Co., Ltd.
**Other Consultants:** BPI (Brandston Partnership, Inc.) (lighting); CS Caulkins Co. Inc (façade maintenance); Edgett Williams Consulting Group Inc. (vertical transportation); SWA Group (landscape); WSP Hong Kong Ltd. (fire)

南京市河西区正在从农村转变,大步迈向繁荣商港和城市中心。金奥国际中心是这片新区的支柱与门户,标志着新兴的都市风貌。这座新一代商业与酒店大楼将性能、效率与住客体验提升到了极致。它独特的外部支撑系统钢材用量较小,节省了20%的整体建筑材料,但却提供了更有效的侧向支撑。

该大楼的设计基于这样一个理念:不仅要将自己与邻近的地块打造成新区门户,也要将其作为这片区域繁荣的象征。它的外立面结构由双层幕墙层层并列而成,这一富有新意的设计在建筑内外层之间形成了遮阳隔热层。最外层幕墙上设置了通风口,能够通过气压将累积的热量从空心墙之间排出,从而进一步降低外墙内侧的温度。

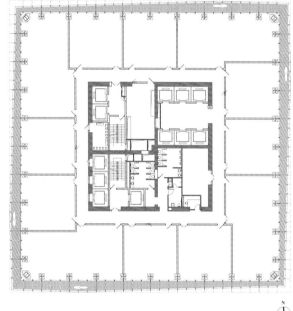

竣工时间:2014 年 9 月
高度:231 m (759 ft)
层数:56
面积:230 621 m² (2 482 384 ft²)
主要功能:酒店 / 办公
业主 / 开发商:江苏金大地(集团)房地产开发有限责任公司
建筑设计:SOM 建筑事务所
结构设计:SOM 建筑事务所
机电设计:科进 | 柏诚集团
总承包商:武汉建工(集团)有限公司
其他顾问方:美国碧谱照明设计有限公司(照明);CS Caulkins 有限公司(外立面维护);Edgett Williams 咨询集团(垂直交通);SWA 集团(景观);WSP 香港有限公司(消防)

**Opposite Left:** Exterior view expressing 'paper lantern' design concept
对侧左图:纸灯笼概念的外立面造型

**Top Left:** Interior view of the entrance lobby
上左图:入口大堂内景

**Top Right:** View looking up the hotel atrium
上右图:仰视中庭

**Bottom:** Typical floor plan
下图:标准层平面图

# China Best Tall Building – Nominee
中国最佳高层建筑奖提名作品

## Oriental Blue Ocean International Plaza
东方蓝海国际广场

Shanghai │ 上海

1700 Huangxing Road was developed out of the desire to create a building that is unique in its use of vibrant color and tactile materials, rather than its height or shape. It is designed to stand out from the monotony of the surrounding grey and glass towers that are ubiquitous to the city. The final built aesthetic is a response to both concept and climate; a visually intriguing facade of blazing red, terra-cotta patterns.

The simple repetitive grid work of clay fins reinforces the solid-looking aesthetic, while providing a sustainable and functional solution, reducing glare in all directions and providing shade from the sun. A series of individually operable windows for natural ventilation are also integrated into the façade. The lobby and retail level contains a dramatic 25-meter-tall entry that incorporates ribbons of black granite with a chiseled stone wall.

**Completion Date:** January 2014
**Height:** 106 m (346 ft)
**Stories:** 22
**Area:** 42,350 sq m (455,852 sq ft)
**Primary Function:** Office
**Owner/Developer:** Shanghai Oriental Blue Ocean Real Estate Co., Ltd.
**Architect:** Perkins + Will
**Structural Engineer:** Zhongxin Architectural Design and Research Institute Pty. Ltd.
**MEP Engineer:** Zhongxin Architectural Design and Research Institute Pty. Ltd.
**Main Contractor:** China State Construction Engineering Corporation
**Other Consultants:** JDC Global Inc. (landscape); ECADI (landscape); Shanghai Songer Lighting Design Co., Ltd. (lighting); Sheng Lue Architectural Technology Co., Ltd. (façade)

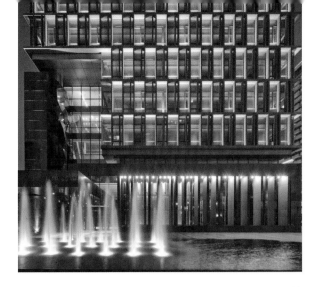

东方蓝海国际广场源于创造一座独特建筑的愿望：其独特之处不在于高度或形状，而在于富有活力和触感的颜色与材料。设计的目标就是让它从单调的灰色背景与城市中无处不在的玻璃幕墙中脱颖而出。外立面很朴素，但对建筑理念与当地气候作出了有力的回应，其闪耀的赤陶图案让外立面显得别具一格。

陶板排列成为简单重复的格栅网络，既加强了外观的坚实感，同时构建一种实用、可持续的眩光解决策略，减弱了各方向的眩光强度，还提供了遮阳的荫凉。外立面上包含一系列用于自然通风的独立活动窗。大堂和商业区有一个高达25 m的醒目入口，由黑色花岗岩条与雕刻石墙构成。

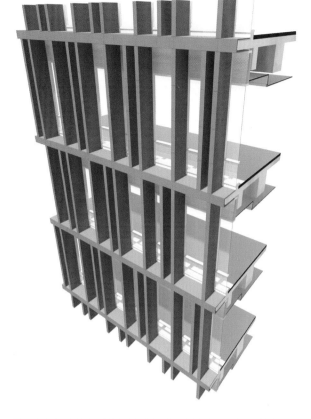

竣工时间：2014 年 1 月
高度：106 m（346 ft）
层数：22
面积：42 350 m² (455 852 ft²)
主要功能：办公
业主/开发商：上海东方蓝海置业有限公司
建筑设计：帕金斯威尔建筑设计事务所（设计）
结构设计：中信建筑设计研究院有限公司
机电设计：中信建筑设计研究院有限公司
总承包商：中国建筑工程总公司
其他顾问方：JDC Global 有限公司（景观）；华东建筑设计研究总院（景观）；上海松尔照明设计有限公司（照明）；盛略建筑科技有限公司（外墙）

**Opposite Left:** Overall view showing terra cotta sleeve
对侧左图：全景
**Top Left:** Building entry
上左图：建筑入口
**Top Right:** Detail façade diagram illustrating terra-cotta fins
上右图：赤陶立面细部
**Bottom Right:** View from southeast into sunken plaza
下右图：从东南向看下沉广场景观

# China Best Tall Building – Nominee
## 中国最佳高层建筑奖提名作品

# Oriental Financial Center | 东方汇经中心
### Shanghai | 上海

The Oriental Financial Center (OFC) is located in the heart of Shanghai's prestigious Lujiazui CBD, enjoying bountiful views of the high-rise cityscape and panoramic views of the Bund across the river. The central outdoor courtyard is the highlight of the building design. The top and bottom of the outdoor courtyard remain completely open, while the traditional building core is split into four separate parts on each corner, unified by a large diagonal steel bracing system. The design is inspired by the double helix pattern of DNA. This motif meets the requirements of structural and mechanical systems, as well as architectural aesthetics.

The central outdoor courtyard design, inspired by the traditional courtyard layout seen in Shanghai's older neighborhoods, increases the use of natural daylight. High-zone office areas near the atrium are fitted with operable windows on the side, providing natural ventilation in spring and autumn. The reduced consumption of artificial lighting and ventilation contributes to a more ecologically sound and pleasant office environment.

**Completion Date:** June 2014
**Height:** 198 m (650 ft)
**Stories:** 31
**Area:** 113,382 sq m (1,220,434 sq ft)
**Use:** Office
**Developer:** Cheung Kong Holdings
**Architect:** Kohn Pedersen Fox Associates; ECADI
**Structural Engineer:** Meinhardt; ECADI
**MEP Engineer:** ECADI
**Main Contractor:** China Construction Eighth Engineering Division Corp. Ltd.
**Other Consultants:** Design Land Collaborative (landscape); Meinhardt (façade); Rolf Jensen & Associates (fire)

东方汇经中心（OFC）位于上海著名的陆家嘴中央商务区中心地段，坐拥城市高层景观的广阔视野，尽享黄浦江两岸的外滩全景。户外中庭是设计的亮点。户外庭院的顶部和底部保持完全的开放，同时传统建筑的核心被分成了四个独立部分，分置于四角，并且通过大型钢架结构支撑系统连接在一起。设计的灵感来源于DNA双螺旋结构。这一主题既符合结构系统和机械系统，又满足了建筑美学的需要。

户外中庭的设计灵感来源于上海的传统庭院布局，它大大提高了自然光的使用率。中庭附近的高层办公区域在一侧安装了活动窗，可以在春秋两季提供自然通风。由此减少的人工照明和通风能耗有助于创造出更自然环保的办公环境。

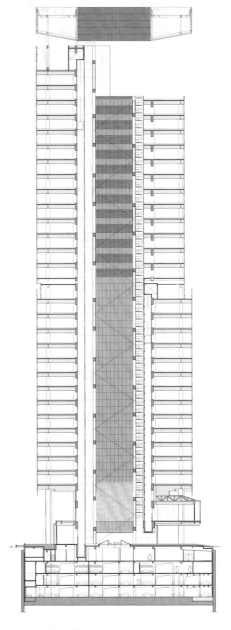

竣工时间：2014年6月
高度：198 m（650 ft）
层数：31
总面积：113 382 m²（1 220 434 ft²）
主要功能：办公
开发商：长江实业
建筑设计：KPF建筑事务所；华东建筑设计研究总院
结构设计：华东建筑设计研究总院；迈进工程设计咨询有限公司
机电设计：巴马丹拿建筑设计咨询（上海）有限公司；华东建筑设计研究总院
总承包商：中国建筑第八工程局有限公司
其他顾问方：地茂景观设计咨询（上海）有限公司（景观）；迈进工程设计咨询有限公司（外立面）；罗尔夫杰森消防技术咨询有限公司（消防）

**Opposite Left:** Overall view
对侧左图：全景
**Top Left:** Interior view of the courtyard lobby
上左图：庭院内景
**Right:** Building section
右图：建筑剖面图

# China Best Tall Building – Nominee
## 中国最佳高层建筑奖提名作品

## R&F Yingkai Square | 富力盈凯广场
### Guangzhou | 广州

Inspired by the segmentation and veining of Chinese bamboo, R&F Yingkai Square is designed as a singular volume that pinches at the corners in relation to the various functions stacked within. This not only gives the tower its unique skyline silhouette, but also signifies at an urban scale how the building is arranged internally.

Varying floor-to-floor heights create an expression of tension and compression on the façade by elongating and shortening the respective floor heights. This is reflected in the design through the use of stainless steel and aluminum vertical strips, which are allowed to stretch and compress as floor heights vary. As the tower meets the ground, the stainless steel increases in density until it creates a completely solid mass in response to the weight of the tower above. Lighting is integrated into the metal tendons and corner carves so that the building's design traits are recognizable in the evening as well.

**Completion Date:** November 2014
**Height:** 296 m (972 ft)
**Stories:** 66
**Area:** 142,956 sq m (1,538,766 sq ft)
**Primary Functions:** Residential / Hotel / Office
**Owner/Developer:** Guangzhou R&F Properties Co., Ltd.
**Architects:** Goettsch Partners; Guangzhou Residential Architectural Design Institute
**Structural Engineer:** Beijing R&F Properties Development Co., Ltd.
**MEP Engineer:** Arup
**Main Contractor:** Guangzhou R&F Properties Co., Ltd.
**Other Consultants:** ACLA (landscape); BPI (lighting); Steve Leung Designers (interiors); Super Potato (interiors)

受中国竹节中空形态的启发，富力盈凯广场独树一帜，在众多功能重叠区的拐角处采用了内缩结构。这一设计不仅塑造出了独特的天际线轮廓，也在城市尺度上表现了内部布局方式。

拉伸和缩短各层层高带来的张力与压力造就了幕墙的独特外观。这在设计中是通过不锈钢和铝合金垂直带反映出来的，它保证了幕墙在层高不同情况下能够稳定拉伸和压缩。越靠近地基，不锈钢的密度越大，由此创造了坚实可靠的基础来承担建筑上部的重量。照明系统、钢筋结构、拐角曲线融合在一起，使大厦的设计特色即便在夜间也清晰可辨。

竣工时间：2014 年 11 月
高度：296 m（972 ft）
层数：66
面积：142 956 m²（1 538 766 ft²）
主要功能：住宅 / 酒店 / 办公
业主 / 开发商：广州富力地产股份有限公司
建筑设计：美国 GP 建筑事务所；广州市住宅建筑设计院
结构设计：北京富力地产股份有限公司
机电设计：英国奥雅纳工程顾问公司
总承包商：广州富力地产股份有限公司
其他顾问方：傲林国际设计有限公司（景观）；美国碧谱照明设计有限公司（照明）；梁志天设计师有限公司（室内设计）；日本 Super Potato 设计事务所（室内设计）

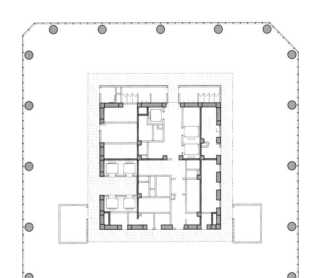

Opposite Left: Overall view
对侧左图：全景
Top Left: Detail view of the façade
上左图：立面细部
Top Right: Typical hotel floor plan
上右图：酒店标准层平面图
Bottom: Typical office floor plan
下图：办公标准层平面图

# China Best Tall Building – Nominee
# 中国最佳高层建筑奖提名作品

## Shanghai Arch | 上海金虹桥国际中心
Shanghai | 上海

The name for the Shanghai Arch mixed-use complex was inspired by the "portal" design of its signature office building. As the first of three components that make up the complex, the office building acts as a gateway into the other project components, as well as to the overall Hongqiao Expansion Business Zone.

Rising up as two separate towers, the building is joined at the 23rd floor by a seven-story skybridge that creates a dynamic "gate" into the pedestrian-friendly retail promenade and offers larger floor plates on the building's upper levels. The east wall is subtly curved with overlapping glass panels, and the east and west elevations feature a double-skin façade for enhanced thermal performance. The glazing at street level blurs the distinction between inside and out, complemented by the flow of the plaza landscaping and water features that "pass through" the office building and connect it to the surrounding context.

**Completion Date:** September 2014
**Height:** 144 m (471 ft)
**Stories:** 29
**Area:** 146,383 sq m (1,575,653 sq ft)
**Use:** Office
**Owner/Developer:** Shanghai Jin Hong Qiao International Property Co., Ltd.
**Architect:** John Portman & Associates; ECADI
**Structural Engineer:** John Portman & Associates; ECADI
**MEP Engineer:** Newcomb & Boyd; ECADI
**Main Contractor:** Shanghai Construction No. 7 (Group) Co., Ltd.
**Other Consultants:** ALT Limited (façade); Arnold Associates (landscape); Fortune Consultants, Ltd. (vertical transportation)

上海金虹桥国际中心综合体的命名灵感来源于其办公大楼的"门户"设计。这座办公大楼是综合体三部分中的第一座建筑,它成为通向项目其他组成部分及整个虹桥商务区拓展区的大门。

两栋楼在低层是分立的,在23层处通过七层高的天桥连接在了一起,形成了通往利于步行的商铺长廊的动态"大门",同时为建筑上层提供了较大的楼层平面。东墙由互相重叠、稍呈弧度的玻璃板构成,同时东西侧的外墙均为能增强保温性能的双层幕墙。地面的临街落地玻璃窗模糊了建筑的内外边界,广场景观与穿越办公楼的水景将建筑与周围环境融为一体。

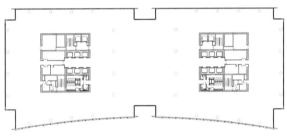
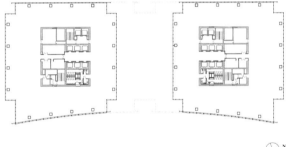

竣工时间: 2014年9月
高度: 144 m (471 ft)
层数: 29
面积: 146 383 m² (1 575 653 ft²)
主要功能: 办公
业主／开发商: 上海金虹桥国际置业有限公司
建筑设计: 约翰·波特曼建筑设计事务所;华东建筑设计研究总院
结构设计: 约翰·波特曼建筑设计事务所;华东建筑设计研究总院
机电设计: Newcomb & Boyd 工程顾问公司;华东建筑设计研究总院
总承包商: 上海建工第七建筑有限公司
其他顾问方: ALT 有限公司(外立面);Arnold 设计公司(景观);财富顾问公司(垂直交通)

**Opposite Top:** Overall view
对侧上图: 全景

**Opposite Bottom:** Interior view of the entrance lobby
对侧下图: 入口大堂内景

**Top Left:** Detail view of the façade's overlapping glass panels
上左图: 玻璃幕墙细部

**Top Right:** View looking up toward the skybridge
上右图: 仰视天桥

**Middle:** Office floor plan on bridge level
中图: 天桥层的办公平面图

**Bottom:** Office floor plan on tower levels
下图: 塔楼的办公平面图

# China Best Tall Building – Nominee
## 中国最佳高层建筑奖提名作品

## Shenzhen Zhongzhou Holdings Financial Center | 深圳中洲控股金融中心
Shenzhen | 深圳

Shenzhen Zhongzhou Holdings Finance Center is a mixed-use complex located at the corner of the Nanshan Culture District in Shenzhen, consisting of a podium, 300-meter office / hotel tower and a 160-meter business apartment tower. With the light evenly distributed on their curved facades, the forms stand in harmony with each other, highlighting their profiles and exhibiting the unique elegance of the composition.

The design is equally appreciable from within. Atria of various scales are placed throughout the building, from double-height areas that create open common spaces, to the towering 17-story hotel lobby. The quality of these spaces is enhanced by the external skin's design elements, which earned the project a LEED Gold rating. Insulated low-E glass has been used extensively. By placing multi-layered horizontal shading panels in crucial areas, diffused light is introduced into indoor spaces, reducing lighting energy consumption, and lowering urban light pollution. Patterned colored glazing in portions of the façade has lowered glare while increasing shading performance.

**Completion Date:** 2014
**Height:** Tower A: 301 m (987 ft); Tower B: 158 m (517 ft)
**Stories:** Tower A: 61; Tower B: 34
**Use:** Tower A: Hotel / Office; Tower B: Residential
**Owner/Developer:** Shenzhen Investment Holdings Co.
**Architect:** Adrian Smith + Gordon Gill Architecture; Beijing Institute of Architectural Design
**Structural Engineer:** Beijing Institute of Architectural Design
**MEP Engineer:** Beijing Institute of Architectural Design
**Main Contractor:** China Railway Construction Engineering Group

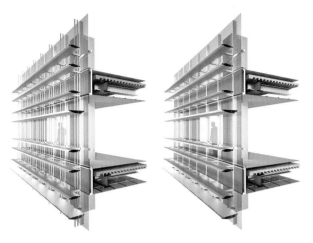

位于深圳市南山文化区的深圳中州控股金融中心是一座多功能综合体，包括一栋裙楼、一栋 300 m 高的集办公、酒店于一体的大楼和一座 160 m 高的商务写字楼。当光线均匀地照射在楼体的曲线上时，各功能区和谐地融为一体，既突出了轮廓，又展示出了建筑独有的优雅。

建筑的室内设计也同样突出。从办公区周围的双倍层高的开放公共空间，到 17 层的酒店大堂，不同体量尺度的空间设计遍布建筑内各处。建筑外墙的设计提高了室内品质，项目也因此获得了 LEED 绿色建筑金牌认证。建筑大量采用了高性能低辐射镀膜玻璃。通过在关键区域安装多层水平遮阳板，漫射光被引入建筑的内部空间，不仅减少了照明系统的能耗，也降低了对城市的光污染。建筑幕墙的压花彩色玻璃在增强遮阳性能的同时，也降低了强光直射。

竣工时间：2014 年
建筑高度：A 座：301 m (987 ft)；B 座：158 m (517 ft)
层数：A 座：61；B 座：34
主要功能：A 座：酒店 / 办公；B 座：住宅
业主 / 开发商：深圳投资控股有限公司
建筑设计：芝加哥 Adrian Smith + Gordon Gill 建筑事务所；北京市建筑设计研究院
结构设计：北京市建筑设计研究院
机电设计：北京市建筑设计研究院
总承包商：中国铁路工程总公司

**Opposite Left:** Overall view from east
对侧左图：东立面全景

**Top Left:** Façade detail showing multi-layered horizontal shading panels
上左图：多层水平向遮阳板的建筑细部

**Top Right:** Interior view looking down the hotel atrium
上右图：鸟瞰中庭内景

**Bottom:** Interior view of office lobby
下图：办公大堂内景

# China Best Tall Building – Nominee
## 中国最佳高层建筑奖提名作品

## Sunrise Kempinski Hotel
### 日出东方凯宾斯基酒店
Beijing | 北京

Although non-traditional in appearance, the Sunrise Kempinski Hotel draws inspiration from a number of traditional Chinese concepts. Located an hour outside of Beijing on the shore of Yanqi Lake, this luxury hotel's most defining characteristic is its rounded form, inspired by the idea of creating a symbiosis with nature through the lens of traditional Chinese philosophy.

To that end, all of the architectural choices in the building reflect some aspect of nature. Most notably, its rounded shape is meant to represent the rising sun. In Chinese culture, sunrise imagery represents vitality, hope, and strength. Symbolizing cooperation and luck, 20 vertical wraps bisect the façade at even intervals. From the side, the building looks like a scallop, representing luck. The entrance building is shaped like the mouth of a fish, representing prosperity. The design concept is more than a symbolic one: The building's floor plates are elliptically shaped, which allows for a 25 percent increase in daylight exposure compared to a conventional box-shaped tower.

**Completion Date:** July 2014
**Height:** 91 m (297 ft)
**Stories:** 20
**Area:** 35,582 sq m (383,001 sq ft)
**Use:** Hotel
**Owner/Developer:** Beijing Enterprises International Conference Metropolis Real Estate Co., Ltd.
**Architect:** Shanghai Huadu Architect Design Company
**Structural Engineer:** Shanghai Huadu Architect Design Company
**MEP Engineer:** Shanghai Huadu Architect Design Company
**Main Contractor:** Beijing Urban Construction Group Co., Ltd

尽管建筑外观看起来并不传统，但日出东方凯宾斯基酒店的设计灵感的确是来源于一系列中国传统元素。酒店坐落于北京雁栖湖南岸，离市区只有一个小时的车程。这座豪华酒店最具标志性的特征便是其圆弧外形，它象征着中国传统哲学中"天人合一"的思想。

为此，酒店的建筑设计无一不体现出"道法自然"的理念。最引人注目的便是那圆形的轮廓，象征着一轮冉冉升起的太阳。它在中国文化里意味着朝气、希望和力量。在外立面上等间距分布有 20 条垂直的龙骨，代表着合作与机遇。从侧面看，酒店形似贝壳，寓意富贵吉祥。主楼入口状如鱼嘴，寓意财源滚滚。当然，酒店独特的设计背后不仅拥有丰富的文化内涵，也有不少实际功能，如由于采用椭圆外形，可使阳光照射比传统的箱式高楼高出 25%。

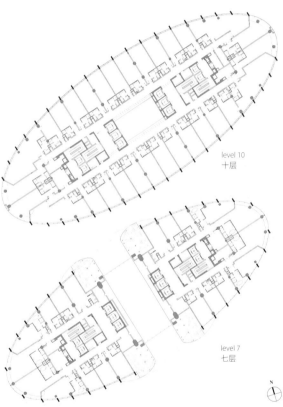

竣工时间：2014 年 7 月
高度：91 m (297 ft)
层数：20
面积：35 582 m² (383 001 ft²)
主要功能：酒店
业主 / 开发商：北京北控国际会都房地产开发有限责任公司
建筑设计：上海华都建筑规划设计有限公司
结构设计：上海华都建筑规划设计有限公司
机电设计：上海华都建筑规划设计有限公司
总承包商：北京城建集团有限责任公司

**Opposite Left:** Overall view from south showing reflection in lake
对侧左图：倒影在湖中的南向全景

**Top Left:** Overall view from southeast
上左图：东南向全景

**Top Right:** Context view from northeast
上右图：东北向全景

**Bottom:** Floor plans: level 10 (top) and level 7 (bottom)
下图：平面图（十层与七层）

# China Best Tall Building – Nominee

## Center 66
## 无锡恒隆广场

Wuxi | 无锡

**Completion Date:** September 2014
**Height:** 250 m (820 ft)
**Stories:** 44
**Area:** 88,560 sq m (953,252 sq ft)
**Use:** Office / Retail
**Owner/Developer:** Hang Lung Properties
**Architect:** Aedas
**Structural Engineer:** Meinhardt
**MEP Engineer:** J. Roger Preston Group
**Main Contractor:** China Construction First Building (Group) Corporation Limited
**Other Consultants:** ALT Limited (façade); Brandston Partnership, Inc. (lighting); Design Land Collaborative (landscape); Rider Levett Bucknall (quantity surveyor)

竣工时间：2014 年 9 月
高度：250 m (820 ft)
层数：44
面积：88 560 m² (953 252 ft²)
主要功能：办公 / 商业零售
业主 / 开发商：恒隆地产
建筑设计：凯达环球
结构设计：迈进工程设计咨询有限公司
机电设计：澧信工程顾问公司
总承包商：中国建筑一局（集团）有限公司
其他顾问方：ALT 国际幕墙顾问公司（外立面）；美国碧谱照明设计有限公司（照明）；地茂景观设计咨询（上海）有限公司（景观）；利比有限公司（工程造价）

## Colorful Yunnan • Flower City
## 七彩云南花之城

Kunming | 昆明

**Completion Date:** January 2015
**Height:** Tower A: 125 m (410 ft); Tower B: 96 m (315 ft)
**Stories:** Tower A: 27; Tower B: 21
**Use:** Hotel / Retail
**Owner/Developer:** Yunnan Mythic Flora Corporate
**Architect:** Tongji Architectural Design (Group) Co., Ltd.
**Structural Engineer:** Tongji Architectural Design (Group) Co., Ltd.
**MEP Engineer:** Tongji Architectural Design (Group) Co., Ltd.

竣工时间：2015 年 1 月
高度：A 座：125 m (410 ft)；B 座：96 m (315 ft)
层数：A 座：27；B 座：21
主要功能：酒店 / 商业零售 / 植物园
业主 / 开发商：云南怡美天香有限公司
建筑设计：同济大学建筑设计研究院（集团）有限公司
结构设计：同济大学建筑设计研究院（集团）有限公司
机电设计：同济大学建筑设计研究院（集团）有限公司

# 中国最佳高层建筑奖提名作品

## Corporate Avenue 6, 7 & 8
## 企业天地 6, 7, 8 号楼
Chongqing | 重庆

**Completion Date:** August 2014
**Height:** 6 Corporate Avenue: 121 m (396 ft); 7 Corporate Avenue: 155 m (508 ft); 8 Corporate Avenue: 180 m (591 ft)
**Stories:** 6 Corporate Avenue: 24; 7 Corporate Avenue: 31; 8 Corporate Avenue: 37
**Use:** Office
**Owner/Developer:** Shui On Land Limited
**Architect:** P & T Group
**Structural Engineer:** P & T Group
**MEP Engineer:** P & T Group
**Main Contractor:** China Construction Eighth Engineering Division Corp. Ltd.
**Other Consultants:** Brandston Partnership, Inc. (lighting); Environmental Market Solutions, Inc. (LEED); Inhabit Group (façade); Rider Levett Bucknall (quantity surveyor)

竣工时间：2014 年 8 月
高度：6 号楼：121 m (396 ft)；7 号楼：155 m (508ft)；8 号楼：180 m (591 ft)
层数：6 号楼：24；7 号楼：31；8 号楼：37
主要功能：办公
业主 / 开发商：瑞安房地产有限公司
建筑设计：巴马丹拿建筑设计咨询有限公司
结构设计：巴马丹拿建筑设计咨询有限公司
机电设计：巴马丹拿建筑设计咨询有限公司
总承包商：中国建筑第八工程局有限公司
其他顾问方：美国碧谱照明设计有限公司（照明）；EMSI（LEED）；英海特工程咨询集团（外立面）；利比有限公司（工程造价）

## Dachong Commercial Center
## 大涌商务中心
Shenzhen | 深圳

**Completion Date:** July 2015
**Height:** Tower 1: 153 m (502 ft); Tower 2: 149 m (490 ft); Tower 3: 130 m (428 ft); Tower 4: 110 m (362 ft)
**Stories:** Tower 1: 35; Tower 2: 34; Tower 3: 29; Tower 4: 24
**Use:** Office
**Owner:** Shenzhen Dachong Industrial Co., Ltd.
**Developer:** China Resources Land Limited
**Architect:** Shenzhen Huayang International Engineering Design Co., Ltd.
**Structural Engineer:** Shenzhen Huayang International Engineering Design Co., Ltd.
**MEP Engineer:** Shenzhen Huayang International Engineering Design Co., Ltd.
**Main Contractor:** China Construction Fourth Engineering Division Corp. Ltd.; China Construction Third Engineering Bureau Co., Ltd.

竣工时间：2015 年 7 月
高度：1 号楼：153 m (502 ft)；2 号楼：149 m (490 ft)；3 号楼：130 m (428 ft)；4 号楼：110 m (362 ft)
层数：1 号楼：35；2 号楼：34；3 号楼：29；4 号楼：24
主要功能：办公
业主：深圳市大冲实业股份有限公司
开发商：华润置地有限公司
建筑设计：深圳市华阳国际工程设计股份有限公司
结构设计：深圳市华阳国际工程设计股份有限公司
机电设计：深圳市华阳国际工程设计股份有限公司
总承包商：中国建筑第四工程局有限公司；中国建筑第三工程局有限公司

# China Best Tall Building – Nominee

## Ding Sheng BHW Taiwan Central Plaza
## 鼎盛BHW台湾中心广场

Taichung | 台中

**Completion Date:** January 2015
**Height:** 159 m (521 ft)
**Stories:** 36
**Area:** 104,138 sq m (1,120,932 sq ft)
**Use:** Office
**Owner/Developer:** High Wealth Construction
**Architect:** EHS ArchiLab + Hsuyuan Kuo Architects & Associates
**Structural Engineer:** Ke Jian Structural Consultants
**MEP Engineer:** Jhih Jhan MEP Consultants
**Main Contractor:** Chyi Yuh Construction
**Other Consultants:** Da Join Architects & Associates (code)

竣工时间：2015 年 1 月
高度：159 m (521 ft)
层数：36
面积：104 138 m² (1 120 932 ft²)
主要功能：办公
业主 / 开发商：兴富发建设
建筑设计：大尺设计 + 郭旭原建筑师事务所
结构设计：Ke Jian 结构顾问公司
机电设计：Jhih Jhan 机电工程顾问公司
总承包商：Chyi Yuh 建筑公司
其他顾问方：Da Join 建筑设计公司（规范）

## Evergrande Huazhi Plaza
## 恒大华置广场

Chengdu | 成都

**Completion Date:** September 2015
**Height:** Office Tower: 192 m (631 ft); Residential Tower 1: 191 m (625 ft); Residential Tower 2: 191 m (625 ft); St. Regis Hotel: 144 m (471 ft)
**Stories:** Office Tower: 36; Residential Tower 1: 53; Residential Tower 2: 53; St. Regis Hotel: 29
**Use:** Office; Residential; Residential; Hotel
**Owner/Developer:** Evergrande Real Estate
**Architect:** Aedas
**Structural Engineer:** AECOM
**MEP Engineer:** WSP | Parsons Brinckerhoff
**Main Contractor:** China State Construction Engineering Corporation
**Other Consultants:** Aurecon (façade); Benoy (interiors); BPI (landscape); China Southwest Design & Research Institute (development consultant)

竣工时间：2015 年 9 月
高度：办公楼：192 m (631 ft)；1 号住宅楼：191 m (625 ft)；2 号住宅：191 m (625 ft)；瑞吉酒店：144 m (471 ft)
层数：办公楼：36；1 号住宅楼：53；2 号住宅楼：53；瑞吉酒店：29
主要功能：办公 / 住宅 / 酒店
业主 / 开发商：恒大地产
建筑设计：凯达环球
结构设计：艾奕康建筑设计有限公司
机电设计：科进 | 柏诚集团
总承包商：中国建筑工程总公司
其他顾问方：澳昱冠工程咨询有限公司（外立面）；英国贝诺建筑设计公司（室内设计）；美国碧谱照明设计有限公司（景观）；中国建筑西南设计研究院有限公司（开发咨询）

# 中国最佳高层建筑奖提名作品

## Fuzhou Shenglong Financial Center
## 福州升龙汇金中心

Fuzhou | 福州

**Completion Date:** December 2014
**Height:** 207 m (678 ft)
**Stories:** 46
**Area:** 156,072 sq m (1,679,945 sq ft)
**Use:** Office
**Owner/Developer:** Shenglong Group
**Architect:** Dennis Lau & Ng Chun Man Architects & Engineers (HK) Ltd. (DLN)
**Structural Engineer:** CCDI Group
**MEP Engineer:** CCDI Group
**Main Contractor:** Henan Guoji Gongcheng Jianshe Gufen Co. Ltd.
**Other Consultants:** Ruihua Construction (façade); Schindler (vertical transportation)

竣工时间：2014 年 12 月
高度：207 m (678 ft)
层数：46
面积：156 072 m² (1 679 945 ft²)
主要功能：办公
业主 / 开发商：升龙集团
建筑设计：刘荣广伍振民建筑师事务所有限公司
结构设计：悉地国际
机电设计：悉地国际
总承包商：河南国基工程建设股份有限公司
其他顾问方：深圳市瑞华建设股份有限公司（外立面）；迅达集团（垂直交通）

## Global Harbor
## 环球港

Shanghai | 上海

**Completion Date:** January 2014
**Height:** 245 m (804 ft)
**Stories:** 45
**Area:** 154,000 sq m (1,657,642 sq ft)
**Use:** North Tower: Serviced apartments / Office / Retail; South Tower: Office / Residential / Hotel / Retail
**Owner/Developer:** Yuexing Group
**Architect:** Chapman Taylor
**Structural Engineer:** China Shipbuilding NDRI Engineering Co., Ltd.
**MEP Engineer:** China Shipbuilding NDRI Engineering Co., Ltd.
**Main Contractor:** Shanghai Construction No.2
**Other Consultants:** Leigh & Orange (interiors); Rolf Jensen & Associates (fire); Theo Kondos (lighting)

竣工时间：2014 年 1 月
高度：245 m (804 ft)
层数：45
面积：154 000 m² (1 657 642 ft²)
主要功能：北楼：酒店式公寓 / 办公 / 商业零售；南楼：办公 / 住宅 / 酒店 / 商业零售
所有人 / 开发商：月星集团
建筑设计：查普门·泰勒建筑设计事务所
结构设计：中船第九设计研究院工程有限公司
机电设计：中船第九设计研究院工程有限公司
总承包商：上海建工二建集团有限公司
其他顾问方：利安顾问有限公司（室内设计）；罗尔夫杰森消防技术咨询有限公司（消防）；Theo Kondos（照明）

# China Best Tall Building – Nominee

## Grand Hyatt Dalian
## 大连君悦酒店
Dalian | 大连

**Completion Date:** August 2014
**Height:** 195 m (641 ft)
**Stories:** 47
**Area:** 79,000 sq m (850,349 sq ft)
**Use:** Residential / Hotel
**Owner/Developer:** China Resources Land Limited
**Architect:** Goettsch Partners; China Architecture Design & Research Group
**Structural Engineer:** RBS Architectural Engineering Design Associates
**MEP Engineer:** Meinhardt
**Main Contractor:** China Construction First Building (Group) Corporation Limited
**Other Consultants:** Celia Chu Design (interiors); Food Service Consultants, Ltd. (food service); LTW Designworks (interiors); Rider Levett Bucknall (quantity surveyor); Tino Kwan Lighting Consultants Ltd. (lighting)

竣工时间：2014 年 8 月
高度： 195 m (641 ft)
层数： 47
面积： 79 000 m² (850 349 ft²)
主要功能：住宅 / 酒店
业主 / 开发商：华润置地有限公司
建筑设计： 美国 GP 建筑设计有限公司；中国建筑设计研究院
结构设计： 广州容柏生建筑结构设计事务所
机电设计： 迈进工程设计咨询有限公司
总承包商： 中国建筑一局（集团）有限公司
其他顾问方：Celia Chu Design（室内设计）；富思餐饮服务顾问有限公司（食品服务）；LTW Designworks（室内设计）；利比有限公司（工程造价）；关永权照明设计有限公司（照明）

## JW Marriott Shenzhen Bao'an
## 深圳前海华侨城JW万豪酒店
Shenzhen | 深圳

**Completion Date:** January 2015
**Height:** 112 m (367 ft)
**Stories:** 23
**Area:** 34,454 sq m (370,860 sq ft)
**Use:** Hotel
**Owner/Developer:** Shenzhen OCT Hotel Real Estate Co., Ltd.
**Architect:** John Portman & Associates; Huasen Architectural & Engineering Designing Consultants Ltd.
**Structural Engineer:** John Portman & Associates; Huasen Architectural & Engineering Designing Consultants Ltd.
**MEP Engineer:** Huasen Architectural & Engineering Designing Consultants Ltd.
**Other Consultants:** Hirsch Bedner Associates (interiors)

竣工时间：2015 年 1 月
高度： 112 m (367 ft)
层数： 23
面积： 34 454 m² (370 860 ft²)
主要功能：酒店
业主 / 开发商：深圳市华侨城酒店置业有限公司
建筑设计： 约翰·波特曼建筑设计事务所；深圳华森建筑与工程设计顾问有限公司
结构设计： 约翰·波特曼建筑设计事务所；深圳华森建筑与工程设计顾问有限公司
机电设计： 深圳华森建筑与工程设计顾问有限公司
其他顾问方：HBA 室内设计事务所（室内设计）

# 中国最佳高层建筑奖提名作品

## Mount Parker Residences
## 西湾台1号

Hong Kong | 香港

**Completion Date:** February 2014
**Height:** 82 m (268 ft)
**Stories:** 21
**Area:** 14,820 sq m (159,521 sq ft)
**Use:** Residential
**Owner/Developer:** Swire Properties
**Architect:** Arquitectonica; Dennis Lau & Ng Chun Man Architects & Engineers (HK) Ltd. (DLN)
**Structural Engineer:** Arup
**MEP Engineer:** J. Roger Preston Group; Meinhardt
**Main Contractor:** Hsin Chong Group
**Other Consultants:** Adrian L. Norman Limited (landscape); Arup (geotechnical); Buro Happold (façade); MVA Transportation, Planning & Management Consultants (traffic)

竣工时间:2014年2月
高度:82 m (268 ft)
层数:21
总占地面积:14 820 m² (159 521 ft²)
主要功能:住宅
业主/开发商:太古地产
建筑设计:Arquitectonica 建筑设计事务所;刘荣广伍振民建筑师事务所有限公司
结构设计:奥雅纳工程咨询有限公司
机电设计:澧信工程顾问公司;迈进工程设计咨询有限公司
总承包商:新昌集团
其他顾问方:Adrian L. Norman 有限公司(景观);奥雅纳工程咨询有限公司(岩土工程);英国标赫工程顾问公司(外立面);MVA 交通规划与管理顾问公司(交通组织)

## Ningbo Global Shipping Plaza
## 宁波环球航运广场

Ningbo | 宁波

**Completion Date:** September 2015
**Height:** 257 m (843 ft)
**Stories:** 52
**Area:** 143,236 sq m (1,541,779 sq ft)
**Use:** Office
**Owner:** Ningbo Global Properties Limited
**Architect:** Nikken Sekkei Ltd; Institute of Shanghai Architectural Design & Research
**Structural Engineer:** Nikken Sekkei Ltd; Institute of Shanghai Architectural Design & Research
**MEP Engineer:** Nikken Sekkei Ltd; Institute of Shanghai Architectural Design & Research
**Other Consultants:** Chuang Qing Facade Consultant (façade); Mindscape Ltd. (landscape)

竣工时间:2015年9月
高度:257 m (843 ft)
层数:52
面积:143 236 m² (1 541 779 ft²)
主要功能:办公
业主:宁波环球置业有限公司
建筑设计:株式会社日建设计;上海建筑设计研究院有限公司
结构设计:株式会社日建设计;上海建筑设计研究院有限公司
机电设计:株式会社日建设计;上海建筑设计研究院有限公司
其他顾问方:创青幕墙顾问公司(外立面);Mindscape 有限公司(景观)

# China Best Tall Building – Nominee

## Shaoxing Shimao Crowne Plaza
## 绍兴世茂皇冠假日酒店

Shaoxing | 绍兴

**Completion Date:** January 2014
**Height:** 288 m (945 ft)
**Stories:** 60
**Area:** 142,714 sq m (1,536,161 sq ft)
**Use:** Hotel / Office
**Owner/Developer:** Shimao Group
**Architect:** ECADI
**Structural Engineer:** ECADI
**MEP Engineer:** ECADI
**Main Contractor:** Shanghai Construction No. 1 (Group) Co., Ltd.

竣工时间：2014 年 1 月
高度：288 m (945 ft)
层数：60
面积：142 714 m² (1 536 161 ft²)
主要功能：酒店 / 办公
业主 / 开发商：世茂集团
建筑设计：华东建筑设计研究总院
结构设计：华东建筑设计研究总院
机电设计：华东建筑设计研究总院
总承包商：上海建工一建集团有限公司

## Shenzhen Xinhe World Office
## 深圳星河World写字楼

Shenzhen | 深圳

**Completion Date:** January 2015
**Height:** 156 m (511 ft)
**Stories:** 31
**Use:** Office
**Owner/Developer:** Shenzhen Yabao Real Estate Development Co., Ltd.
**Architect:** AECOM
**Structural Engineer:** AECOM
**MEP Engineer:** AECOM
**Main Contractor:** China Construction First Building (Group) Corporation Limited

竣工时间：2015 年 1 月
高度：156 m (511 ft)
层数：31
主要功能：办公
业主 / 开发商：深圳市雅宝房地产开发有限公司
建筑设计：艾奕康建筑设计有限公司
结构设计：艾奕康建筑设计有限公司
机电设计：艾奕康建筑设计有限公司
总承包商：中国建筑一局（集团）有限公司

# 中国最佳高层建筑奖提名作品

## Studio City
## 新濠影汇

Macau | 澳门

**Completion Date:** October 2015
**Height:** 152 m (499 ft)
**Stories:** 32
**Area:** 160,000 sq m (1,722,226 sq ft)
**Use:** Hotel / Casino
**Owner/Developer:** Melco Crown Entertainment Limited
**Architect:** Goddard Group; Leigh & Orange
**Structural Engineer:** AECOM
**MEP Engineer:** Meinhardt
**Main Contractor:** Paul Y. Engineering Group; Yau Lee Group
**Other Consultants:** Arup (fire); Earth Asia (landscape); Illumination Physics (lighting); Inhabit Group (façade); LTW Designworks (interiors); MVA Transportation, Planning & Management Consultants (traffic); Shen Milsom Wilke, Inc. (acoustics); Tino Kwan Lighting Consultants Ltd. (lighting); WT Partnership (quantity surveyor)

竣工时间：2015 年 10 月
高度：152 m (499 ft)
层数：32
面积：160 000 m² (1 722 226 ft²)
主要功能：酒店 / 博彩
业主 / 开发商：新濠博亚娱乐有限公司
建筑设计：Goddard 设计集团；利安顾问有限公司
结构设计：AECOM
机电设计：迈进工程设计咨询有限公司
总承包商：保华建业；有利集团
其他顾问方：奥雅纳工程咨询有限公司（消防）；泛亚国际（景观）；Illumination Physics（照明）；英海特工程咨询集团（外立面）；LTW Designworks（室内设计）；MVA 弘达交通规划与管理顾问公司（交通组织）；声美华顾问公司（声学）；关永权照明设计有限公司（照明）；务腾咨询有限公司（工程造价）

## The Wave of Science and Technology Park S01
## 浪潮科技园S01科研楼

Ji'nan | 济南

**Completion Date:** August 2014
**Height:** 161 m (529 ft)
**Stories:** 35
**Area:** 114,452 sq m (1,231,951 sq ft)
**Use:** Office
**Owner:** Inspur Group
**Developer:** Ji'nan Inspur Mingda Information Technology Co., Ltd.
**Architect:** Beijing Institute of Architectural Design
**Structural Engineer:** Beijing Institute of Architectural Design
**MEP Engineer:** Beijing Institute of Architectural Design
**Main Contractor:** Tianyuan Construction Group Co., Ltd.

竣工时间：2014 年 8 月
高度：161 m (529 ft)
层数：35
面积：114 452 m² (1 231 951 ft²)
主要功能：办公
业主：浪潮集团
开发商：济南浪潮铭达信息科技有限公司
建筑设计：北京市建筑设计研究院
结构设计：北京市建筑设计研究院
机电设计：北京市建筑设计研究院
总承包商：天元建设集团

# China Best Tall Building – Nominee

## Tianjin International Trade Tower 1, 2 & 3
## 天津国际贸易中心1, 2, 3号楼

Tianjin | 天津

**Completion Date:** October 2014
**Height:** Tower 1: 235 m (771 ft); Tower 2: 165 m (542 ft); Tower 3: 165 m (542 ft)
**Stories:** Tower 1: 57; Tower 2: 41; Tower 3: 45
**Area:** 155,000 sq m (1,668,406 sq ft)
**Use:** Tower 1: Serviced apartments / Retail; Tower 2: Office; Tower 3: Serviced apartments
**Owner/Developer:** CapitaLand Limited
**Architect:** P & T Group
**Structural Engineer:** P & T Group
**MEP Engineer:** Arup; Tianjin Architecture Design Institute
**Main Contractor:** China Construction Third Engineering Bureau Co., Ltd.
**Other Consultants:** Environmental Market Solutions, Inc. (LEED); GDIL Lighting Design (lighting); Inhabit Group (façade); L&A Urban Planning and Landscape Design (Canada) Ltd. (landscape); Rider Levett Bucknall (quantity surveyor); Spark Architects (interiors)

竣工时间：2014 年 10 月
高度：1号楼：235 m (771 ft)；2号楼：165 m (542 ft)；3号楼：165 m (542 ft)
层数：1号楼：57；2号楼：41；3号楼：45
面积：155 000 m² (1 668 406 ft²)
主要功能：1号楼：酒店式公寓 / 商业零售；2号楼：办公；3号楼：酒店式公寓
业主 / 开发商：凯德置地
建筑设计：巴马丹拿建筑设计咨询有限公司
结构设计：巴马丹拿建筑设计咨询有限公司
机电设计：奥雅纳工程咨询有限公司；天津市建筑设计院
总承包商：中国建筑第三工程局有限公司
其他顾问方：EMSI（LEED）；香港大观国际设计咨询有限公司（照明）；英海特工程咨询集团（外立面）；加拿大奥雅景观规划设计事务所（景观）；利比有限公司（工程造价）；思邦建筑（室内设计）

## WPP Campus
## 达邦协作广场

Shanghai | 上海

**Completion Date:** May 2015
**Height:** 150 m (491 ft)
**Stories:** 31
**Area:** 51,120 sq m (550,251 sq ft)
**Use:** Office
**Owner/Developer:** B.M. Holding (Group) Co., Ltd.; Nan Fung Group
**Architect:** Gensler; Shanghai Xian Dai Architecture Design (Group) Co., Ltd.
**Structural Engineer:** Shanghai Xian Dai Architecture Design (Group) Co., Ltd.
**MEP Engineer:** Shanghai Xian Dai Architecture Design (Group) Co., Ltd.
**Main Contractor:** Shanghai Construction Group
**Other Consultants:** HASSELL (interiors, landscape); ISSEY (façade); Meinhardt (façade)

竣工时间：2015 年 5 月
高度：150 m (491 ft)
层数：31
总占地面积：51 120 m² (550 251 ft²)
主要功能：办公
业主 / 开发商：宝矿控股（集团）有限公司；南丰集团
建筑设计：晋思；上海现代建筑设计集团有限公司
结构设计：上海现代建筑设计集团有限公司
机电设计：上海现代建筑设计集团有限公司
总承包商：上海建工集团
其他顾问方：HASSELL 设计公司（室内设计，景观）；ISSEY（外立面）；迈进工程设计咨询有限公司（外立面）

# 中国最佳高层建筑奖提名作品

## Wuxi Suning Plaza 1
## 无锡苏宁广场

Wuxi | 无锡

**Completion Date:** January 2014
**Height:** 328 m (1,076 ft)
**Stories:** 67
**Area:** 116,580 sq m (1,254,857 sq ft)
**Use:** Hotel / Serviced apartments / Office
**Owner/Developer:** Suning Real Estate Group
**Architect:** RTKL
**Structural Engineer:** Jiangsu Provincial Architectural D&R Institute Ltd.
**MEP Engineer:** Jiangsu Provincial Architectural D&R Institute Ltd.
**Main Contractor:** China Construction First Group Construction & Development Co., Ltd.

竣工时间：2014 年 1 月
高度：328 m（1 076 ft）
层数：67
面积：116 580 m²（1 254 857 ft²）
主要功能：酒店 / 酒店式公寓 / 办公
业主 / 开发商：苏宁地产集团
建筑设计：RTKL 建筑事务所
结构设计：江苏省建筑设计研究院有限公司
机电设计：江苏省建筑设计研究院有限公司
总承包商：中建一局集团建设发展有限公司

## Xiamen World Overseas Chinese International Conference Center
## 厦门世侨中心

Xiamen | 厦门

**Completion Date:** December 2014
**Height:** 113 m (370 ft)
**Stories:** 23
**Area:** 45,553 sq m (490,328 sq ft)
**Use:** Office
**Owner:** Jiang Architects and Engineers
**Developer:** WOCICC, Ltd.
**Architect:** Jiang Architects and Engineers
**Structural Engineer:** Jiang Architects and Engineers
**MEP Engineer:** Jiang Architects and Engineers
**Main Contractor:** China State Construction Engineering Corporation

竣工时间：2014 年 12 月
高度：113 m（370 ft）
层数：23
面积：45 553 m²（490 328 ft²）
主要功能：办公
开发商：厦门世侨投资管理有限公司
建筑设计：江欢成建筑设计有限公司
结构设计：江欢成建筑设计有限公司
机电设计：江欢成建筑设计有限公司
总承包商：中国建筑工程总公司

# China Tall Building Urban Habitat Award
中国高层建筑城市人居奖

# Awards Criteria
评选标准

This award acknowledges projects ranging from brilliantly executed master plans or urban designs that have led to a quality urban environment, at the scale of a single site or several sites, where the interface between a tall building and the urban realm is exemplary. Projects should demonstrate a positive contribution to the surrounding environment, add to the social sustainability of both their immediate and wider settings, and represent design influenced by context, both environmental and cultural.

1. The project must be physically located in the Greater China region, including Hong Kong, Macau, Taiwan and Mainland China.
2. Submissions can range in scale from an urban complex on a single site to a community, city comprehensive functional zone, or a complete master plan or urban design of a neighborhood or city, and shall include at least one tall building. Projects with buildings no higher than 100 meters are unlikely to qualify.
3. The "urban habitat" aspect of the project must have been completed and utilized in the two years prior to the current awards year, to have allowed time for the urban habitat to have established itself (e.g., for the 2016 awards, a project must have a completion date between January 1, 2013 and December 14, 2015. Proposals or visions are not eligible. In the case of a master plan or urban design, a multi-phase plan that is only partially completed will be accepted, but the completed portion must be far enough progressed that its urbanistic vision is evident. This completed portion of the project, or the entire project when a complete project is submitted, shall have been completed and approved by related departments and will have been constructed by December 14, 2015. The implementation certification documents shall be provided by the owners.

4. Projects that are submitted for the China Best Tall Building awards are also eligible to submit to the Urban Habitat award in the same year. Projects that have been submitted to the China Best Tall Building award in the previous year are eligible for submission for the Urban Habitat award in the subsequent year, as long as completion has occurred within the dates as outlined above.

该奖项是为了表彰那些创造出高品质城市环境的总体规划或城市设计，或是在单体建筑或建筑群中，使高层建筑与周边城区衔接堪称典范的项目。项目应体现出对周边环境的积极贡献，并强化更长远的社会可持续性，同时还展现出设计所受到的当地环境和文化的影响。

1. 该项目的地点应在大中华地区，包括中国大陆和香港、澳门、台湾地区。
2. 项目规模包括从单一体量的城市综合体项目直至社区、城市综合功能区、城市或社区的完整总体规划或城市设计，并应包含至少一栋100 m以上高层建筑，若项目所含建筑中无100 m以上的高层建筑，将不具备参选资格。
3. 参选的项目必须在提交申请和颁奖年份之前的两年内竣工并投入使用，以便有时间建立起自身的城市人居环境（例如，对于2016年度的奖项，参选项目应在2013年1月1日至2015年12月14日期间竣工）；如仅是项目方案或愿景则不具备参评资格。如果是一项包含若干阶段的总体规划或城市设计，那么该方案也是有资格提交申请的，但完成的体现城市人居理念部分必须有足够的进展，能够体现城市规划师的意图。提交部分完成或全部完成的项目，需在2015年12月14日前完成并通过相关部门审批，已经开始实施建设，申报时要求业主提供相应的实施证明文件。
4. 申报"中国最佳高层建筑奖"评选的项目也可在同一年同时申报"城市人居奖"。申报过"中国最佳高层建筑奖"的项目，只要在上述规定日期内完成，也可申报下一年的"城市人居奖"评选。

# China Tall Building Urban Habitat Award – Winner
中国高层建筑城市人居奖

## Jing An Kerry Center | 静安嘉里中心
Shanghai | 上海

Jing An Kerry Center occupies a prominent location in the Jing An district, which, along with Shanghai overall, has developed swiftly over the past decade. The city's rapid development has led to an increase in the construction of massive residential complexes that are designed to accommodate the maximum number of inhabitants. These monolithic towers suffer at the human scale, and have taken away from the livability of the ancient Jing An district, which

静安嘉里中心位于上海市静安区的黄金地段,静安区近年来正随着上海的整体发展日新月异,城市的高歌猛进带来了大型住宅区的繁荣,这些庞大的建筑群往往以最大限度地容纳住户为目标。静安区自古以来就是上海的文化中心、住宅中心和商业中心,而这些高楼大厦却往往忽视了人性化的考量,带走了静安区的静谧与古香。

有鉴于此,静安嘉里中心的设计专注于创造出空间上没有闭塞感,功能上开放灵活的建筑区,既满足当地日益

**Completion Date:** January 2013
**Total Land Area:** 45,857 sq m (493,601 sq ft)
**Total Building Footprint:** 19,713 sq m (212,189 sq ft)
**Building Height:** Tower 1: 133 m (436 ft); Tower 2: 260 m (853 ft); Tower 3: 198 m (649 ft)
**Primary Function:** Hotel / Office
**Owner/Developer:** Kerry Properties Ltd
**Urban Planner:** Kohn Pedersen Fox Associates
**Architects:** Kohn Pedersen Fox Associates; Wong & Ouyang; Benoy; Hirsch Bedner Associates; Institute of Shanghai Architectural Design & Research; Super Potato
**Landscape Architect:** SWA Group
**Structural Engineer:** Arup
**MEP Engineers:** WSP | Parsons Brinckerhoff; Shanghai Institute of Architectural Design & Research
**Main Contractor:** Shanghai Construction Group

竣工时间:2013 年 1 月
总用地面积:45 857 m² (493 601 ft²)
建筑基底面积:19 713 m² (212 189 ft²)
建筑高度:塔 1:133 m (436 ft);塔 2:260 m (853 ft);塔 3:198 m (649 ft)
主要功能:酒店 / 办公
业主 / 开发者:嘉里建设有限公司
规划设计:KPF 建筑事务所
建筑设计:KPF 建筑事务所;王欧阳香港有限公司;英国贝诺建筑设计公司;HBA 设计工程顾问有限公司;上海建筑设计研究院有限公司;Super Potato
景观设计:SWA 集团
结构设计:Arup
机电设计:科进 | 柏诚;上海建筑设计研究院有限公司
总承包商:上海建工集团

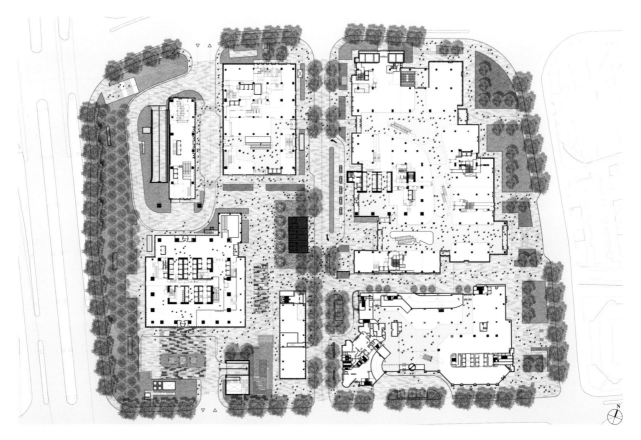

is the historic heart of Shanghai, and the city's cultural, residential, and commercial center.

As a response to this, the design of the Jing An Kerry Center is focused on creating an accessible space and incorporating flexibility in its functions, which can serve to accommodate the area's growing population, and meet the needs of residents and users. The development thus serves multiple functions: housing, retail, transit, and offices, all in one complex designed to accommodate the city's growing population, but built to human scale. The scheme is divided into distinct blocks that correspond to their different functions, with a large plaza to tie it together.

The project site's locale is of historical significance in that it contains one of Mao Zedong's early residences. What at first seems to be one of the most incongruous buildings on the site actually rounds out the complete story, and provides the most strident illustration of China's transformation into a global market center in just a few decades. Located in the center of the development, the residence operates as an anchor for the surrounding buildings, informing their location and layout. The residence has been turned into a museum that enriches the cultural value of the development and attracts more diverse elements to the site.

Along with the historical value of the Mao residence, the development brings a great deal of modern convenience to the neighborhood, and is pedestrian-friendly and highly accessible. It takes advantage of the transportation networks on-site, connecting to Shanghai Metro lines 2 and 7, which converge directly underneath, drawing a combined 2.5 million riders per day. This convergence also translates logically and efficiently into a separate underground pedestrian level that connects the differentiated blocks with unified wayfinding, retail opportunities, and of course, protection from inclement weather. Major vehicular and pedestrian arteries are located on Nanjing Road

增长的人口数量，也充分考虑住户和用户的需要。该建筑区集成了多种功能：居住、零售、交通、办公，在满足城市人口增长的同时又兼顾人性化设计。基于以上指导思想，整个建筑区被划分为由大型广场连接起来的不同功能区。

毛泽东故居坐落于项目区域内，有着极高的历史价值。故居乍看与周边环境格格不入，但实际上它正述说着历史，见证着数十年来中国跻身世界中心的成功转型。这座房子位于建筑区中心，是周围建筑的锚点，标示着他们的地理位置和布局。如今它被改造为博物馆，不仅提高了整个建筑区的文化价值，也吸引了更加多元的要素。

除了毛泽东故居的历史价值之外，这一建筑区对街坊来说也不失现代化的便利，适宜步行且交通畅达。建筑区利用了现有的交通网络，选址于上海地铁 2 号线和 7 号线的交界处，可以从地下直接进入地铁，每天人流量高达 250 万人次。在地块的更新中还成功连接了不同街区的独立地下步行街，可以提供指路标，进行零售，恶劣天气时为行人遮风挡雨。此外，南京西路上还有车辆和人行主干道。

**Previous Spread | 前页**
**Left:** Aerial view of central courtyard space
左图：中轴线广场鸟瞰

**Right:** Courtyard space with preserved Mao Zedong's house on right
右图：广场空间，右侧是毛泽东故居

**Current Spread | 本页**
**Opposite Top:** Site plan, showing ground floor plans, Mao's house depicted in red
对侧上图：总平面图，红色部分为毛泽东故居

**Top Right:** Building entrance, showing how the tower's mass is broken down to a pedestrian scale at its base
上右图：建筑入口，显示出建筑体量如何在建筑底部分解到步行尺度

## Jury Statement | 评委会评语

The domineering scale and persistent rush of China's biggest city are tamed and channeled in a massive development that easily could have exacerbated those conditions. Instead, Jing An Kerry Center is an object lesson in modulation, such that the towers that excite on the skyline have a calming effect by the time they reach the ground. This is sublimely reinforced by both the permeability of the site and the incorporation of small-scale structures, both old and new. Even the underground spaces have seemingly been designed to provide connectivity with an atmosphere of serenity.

作为中国最大的城市之一，上海的规模显得霸气傲然同时带有城市惯有的喧嚣繁忙，大型项目的开发也极易落入这一窠臼。与之相反，静安嘉里中心成功地控制了这个问题。从项目落地开始，它就为附近的天际线增添了一抹宁静的色彩，为都市肌理的调整提供了一个典范。通过强调建筑的通透性以及与新旧小型建筑的融合，这种感觉更加得以彰显。甚至地下空间的设计在提供交通功能的同时，也营造出一种宁静的气氛。

"In many ways, Jing An Kerry Center provides the essential transition between the international high-street glamour of Nanjing Road and the fine-grained lane houses that persist south of the site."

"南京路上有着国际化高楼大厦的风采，而南面则存留着密集排布的居住区，很大程度上来说，静安嘉里中心提供了上述两者间必要的过渡。"

Design Juror　设计评委

West, making the site, located halfway between the Jing An Temple and the Shanghai Exhibition Center, a prime location to infuse with social and commercial activity.

The planning typology recalls Concession-era Shanghai, when the French occupied part of the city in 1849–1943, and it was common to find inner-block lanes, pathways, and courtyards. This allows more of the surrounding urban fabric of streets and walkways to enter freely into the site. It also softens the scale of the larger towers, providing a gradual transition between the human scale of the plaza and the very top of the complex, in a way that is also reminiscent of New York's Rockefeller Center but particular to the Shanghai local condition.

The Jing An Kerry Center has been crafted to be flexible for a variety of tenants, as demand changes. The complex consists of versatile spaces that can easily be repurposed. By dividing the structure into separate elements and avoiding one large-scale form, the individual pieces created are all more adaptable to future uses as the economy changes and the city grows.

On the whole, the complex stimulates commerce and boosts tourism, which generates economic growth locally, and will likely encourage more productive development in the area. With the Mao Zedong residence museum as a centerpiece, the complex is contextualized historically, and a balance is achieved between the consumer outlets on-site and the cultural and educational value that is introduced by the museum. The entire Jing An Kerry Center complex is certified LEED Gold.

*"In addition to its elegant design, Jing An Kerry Center is particularly successful in creating a lively outdoor public place for families, a rare feat in Shanghai."*

"除了设计上的优雅,静安嘉里中心还极其成功地创造了有活力的家庭户外活动区,这在上海实属罕见。"

Design Juror　设计评委

**Opposite Left:** Courtyard with interwoven pedestrian spaces
对侧左图: 庭院广场与交织的步行空间

**Bottom:** Overall view from west
下图: 西面全景

这些都使得位于静安寺和上海展览中心之间的黄金地段成为社会活动和商业活动的最佳选择。

这种规划设计唤起了人们1849年至1943年间上海法租界时代的记忆,区内街道、小径、庭院遍布,这使得周遭车道和人行道的结构能自由融入,高楼的棱角也显得更加柔和,在广场的人性化尺度与建筑群的高度之间创造了自然的过渡。

考虑到形形色色的租户,静安嘉里中心由可调整用途的多功能区组成。通过结构细分,避免规模过大,这些独立区块将在经济发展和城市扩张的进程中更好地适应未来。

总而言之,该建筑群能够刺激贸易流通和旅游业繁荣,促进当地经济增长,甚至能推动生产力的发展。毛泽东故居带来了浓厚的历史气息,实现了消费网点和文化教育价值的平衡。静安嘉里中心建筑群还获得了LEED金奖。

# China Tall Building Urban Habitat Award – Honorable Distinction
中国高层建筑城市人居荣誉奖

## Heart of Lake | 万科湖心岛
Xiamen | 厦门

This 186,000 square-meter pedestrian garden city features a diverse mix of residential types, including high-rise and mid-rise apartments, townhouses, and single-family villas, on a 10-hectare site on Huxindao Island in Xiamen. The residences are organized around a lush green central park, intimate courtyards, and stone-paved plazas; landscaped pedestrian streets lead to a waterfront public park that wraps the peninsula. A highly-visible clubhouse serves as a gateway and social center. As opposed to much recent residential development in China, which relies on slab towers and single-family houses, the design introduces point towers and low-rise and mid-rise apartment buildings, villas, and houses to the mix in Heart of Lake.

Heart of Lake's architectural vocabulary, inspired by both traditional Chinese and Mediterranean architecture, responds to local climate and takes advantage of local materials and construction techniques. The project combines a sense of order

这座步行花园城位于厦门湖心岛，占地 10 hm²，建筑面积 186 000 m²，集合了高层公寓、多层公寓、联排别墅和独栋别墅等多种住宅类型。住宅区附近建有绿草成荫的中央公园、宁静怡人的庭院和铺石广场，还有风景秀丽的步行街通往环绕半岛的临水公园。这里的娱乐中心是入口处引人注目的俱乐部。不同于近年来国内流行的板式高层和独栋别墅，湖心岛设计将高层塔楼、中低层公寓、独栋别墅和联排别墅融为一体。

湖心岛一词源于中国传统建筑风格和地中海风格设计理念，充分考虑当地气候条件，同时也利用了当地的建筑材料和技术。这个项目将有序感与惊喜感结合起来，按照不同等级的街道网络来组织区域，又对某些局部进行了特别设计。本土化与一体化的建筑语言，统一性与多样性的精致结合，不同体量的建筑物求同存异，共同为住户提供了个人空间和社群氛围。项目所需的花岗岩均来自当地。该设计成功将中国城市城墙环绕、界限分明的传统发扬光大。自街道下延至公园的台阶将半岛环绕起来，将高档住宅区与水滨公园联结起来。

**Completion Date:** August 2013
**Total Land Area:** 95,098 sq m (1,023,626 sq ft)
**Total Building Footprint:** 301,846 sq m (3,249,043 sq ft)
**Total Open Area:** 63,240 sq m (680,710 sq ft)
**Total Hardscape:** 29,956 sq m (322,444 sq ft)
**Total Softscape:** 33,284 sq m (358,266 sq ft)
**Primary Function:** residential
**Developer:** Vanke Group
**Urban Planner:** Robert A.M. Stern Architects
**Architects:** Robert A.M. Stern Architects; Xiamen BIAD Architectural Design Ltd.
**Landscape Architect:** Olin Studio
**Structural Engineer:** Xiamen BIAD Architectural Design Ltd.
**MEP Engineer:** Xiamen BIAD Architectural Design Ltd.
**Main Contractor:** China State Construction Fourth Engineering Division Corporation Ltd.

竣工时间：2013 年 8 月
总用地面积：95 098 m²（1 023 626 ft²）
总建筑基底面积：301 846 m²（3 249 043 ft²）
总户外面积：63 240 m²（680 710 ft²）
总硬质景观面积：29 956 m²（322 444 ft²）
总软质景观面积：33 284 m²（358 266 ft²）
主要功能：住宅
开发商：万科集团
规划设计：罗伯特斯特恩建筑师事务所
建筑设计：罗伯特斯特恩建筑师事务所；厦门佰地建筑设计有限公司
结构设计：厦门佰地建筑设计有限公司
机电设计：厦门佰地建筑设计有限公司
总承包商：中国建筑第四工程局有限公司
其他顾问方：Olin Studio（景观设计）

– the whole is organized by a gridded hierarchy of streets—and a sense of surprise, utilizing a kit of parts. Consistent vernacular architectural expression and detailing establish unity within variety, allowing buildings of various scales to enjoy a common character, while providing residents a sense of individuality within a community. Locally-sourced granite was used throughout the project. The design successfully carries forward the tradition of Chinese walled cities with clear edges, but instead, streets open to stairs that descend to the public gardens that wrap the peninsula, creating a connection between the high-end residential neighborhood and the public waterfront park.

Inspired by Gulangyu Island, a nearby historic European Treaty Port that includes a mix of public and private spaces, highly-trafficked markets, and a balance of buildings and streets, Heart of Lake's residences are organized around a hierarchy of landscaped public spaces, a lush green central park, intimate courtyards, and stone-paved plazas. The placement of the buildings on Huxindao Island was also considered in relation to strict sunlight requirements. Natural light in all living spaces and gardens was maximized by locating most of the towers to the east side of the site, and staggering them to reduce the impact of long afternoon shadows.

Heart of Lake is a fully pedestrian development. Public parks are arranged to capture breezes from the north to combat the heat and humidity of the local climate. The neighborhood is built adjacent to an elevated urban transit system, which currently provides bus service and will in the future provide train service. All parking and services are located below the platform that constitutes the community's new ground level, which is raised on reinforced concrete waffle-slabs with pockets that accommodate landscaping. Residents arriving by car access the street level via elevators and stair towers that also serve as light wells for the parking.

## Jury Statement | 评委会评语

Heart of Lake represents the realization of a vision to create a "neighborhood," as opposed to simply a "development." Variations in massing, scale and texture, and the incorporation of pathways and gardens that encourage exploration and contemplation, imbue the project with the potential to create memorable stories and long-lived attachments.

湖心岛的开发创造了一个"社区",而不仅是简单的一个"建筑项目"。通过体量、规模和结构的变化,以及通道和花园的整合,湖心岛创造了一个鼓励探索和沉思的社区,让这里有可能发生令人难忘的故事,产生长久的依恋之情。

> *"Breaking from the tendency toward shiny iconography that characterizes new-build projects, Heart of Lake impels the visitor to feel as if it has always been there."*
>
> "新的建筑项目往往有着炫亮的意态，但湖心岛却逆于潮流，使得游客们有种'它自古以来一直都在'的错觉。"
>
> Design Juror　设计评委

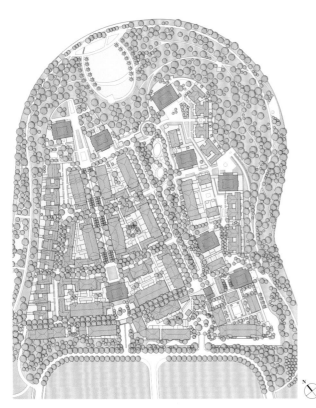

附近的鼓浪屿早在《南京条约》时就被辟为通商口岸，其规划公共空间和私人空间并重，市场人流如织，建筑和街道尺度宜人。受其启发，湖心岛住宅区周围公共空间层级分明，中央公园绿草成荫，还有私密的庭院和石头铺砌的广场。湖心岛建筑的格局也充分考虑了日照需求，大部分高层建筑都安排在东边，并且尽量交错排布，避免午后阴影，使所有住宅区和花园都能尽可能地接收到阳光。

湖心岛是一个完美的散步区。公园可以充分利用从北边吹来的凉风来应对当地湿热的气候。建筑区临近城市高架运输系统，现在已经有公交服务，未来还将开通火车。第一层平台由钢筋混凝土筑成，预留了景观设计的空间，所有停车位都位于平台之下。有私家车的住户可以通过电梯或者楼梯直达地面，楼梯同时也可充当车库的采光井。

**Previous Spread | 前页**

**Left:** Major circulation axes are marked by landscaped pedestrian greensward
左图：交通轴线是一条绿色步行道

**Current Spread | 本页**

**Opposite Top Left:** View of plaza embellished with a fountain
对侧上图：喷泉广场

**Opposite Bottom:** Pedestrian path in one of the courtyards
对侧下图：庭院步道

**Top Left:** Master plan
上左图：总平面图

**Bottom Left:** The clubhouse, which serves as a platform for a neighborhood of townhouses and apartments above
下左图：会所是社区邻里的服务平台

# China Tall Building Urban Habitat Award – Honorable Distinction
中国高层建筑城市人居荣誉奖

## Shenye Tairan Building | 深业泰然大厦
Shenzhen | 深圳

As a featured building containing the offices of a developer of the Chegongmiao District, on a site that occupies an entire city block, the Shenye Tairan Building needed to be inviting on a local, human scale, yet prominent at an urban scale. While the default response to the site's zoning would have been two separate 100-meter towers, set back from the street by a commercial podium, instead the building is conceived as one continuous volume along the street frontage, with the southwest corner built to its maximum allowable height. In response, the opposite northeast corner is scaled down and lifted up from the ground to invite pedestrians into the internal public spaces.

The result is a city block building with a vertical courtyard of treasures - a central communal space framed by the tall mass of the building on two sides, so that occupants feel protected from the urban bustle outside and shaded from the hot southern sun. At the heart of the courtyard, a reflecting pool

作为深圳车公庙地区的特色建筑，深业泰然大厦占据了整个街区，因此它在兼顾当地人文的同时，也要突出都市气度。这一区域最初的规划是分别修建两座 100 m 高的大楼，通过一个商业平台和街道隔开。现在深业泰然大厦的构思则是沿着临街面绵延伸展，西南角尽可能地推高，东北角则压低，从地平面逐步推升，形成一种邀请行人进入的视觉效果。

这样的设计使得街区看起来仿佛珍宝阁一般——两侧是大体量高层建筑，中心存在共享空间，让居住者远离城市的喧嚣，躲避炽热的艳阳。庭院中心建有映景明池，如水晶般为地下车库的中心入口提供亮光。倒影池内还有从内部庭院直接进入大楼的通道，住户们再也不用进入阴暗的停车场、走进隐蔽的电梯才能到达大楼了，这再一次改造了进入办公区的传统方式。

传统办公大楼的屋顶不对公众或住户开放，而深业泰然大厦则利用倾斜的设计，在屋顶建造了开放式楼梯，可到达多个楼层。屋顶构成了建筑的"第五立面"，形成了和

**Completion Date:** December 2012
**Total Land Area:** 24,522 sq m (263,952 sq ft)
**Total Building Footprint:** 10,691 sq m (115,077 sq ft)
**Total Open Area:** 13,832 sq m (148,886 sq ft)
**Building Height:** 106 m (348 ft)
**Primary Function:** Office
**Developer:** Shenzhen Terra (Holdings) Co., Ltd.
**Urban Planner:** Zhubo Design Group Co., Ltd.
**Architect:** Zhubo Design
**Landscape Architects:** Zhubo Design Group Co., Ltd.; Hill Landscape Design
**Structural Engineer:** Zhubo Design Group Co., Ltd.
**MEP Engineers:** Zhubo Design Group Co., Ltd.; China Academy of Building Research
**Main Contractor:** Shenzhen Terra (Holdings) Co., Ltd.

竣工时间：2012 年 12 月
总用地面积：24 522 m² (263 952 ft²)
总建筑基底面积：10 691 m² (115 077 ft²)
总户外面积：13 832 m² (148 886 ft²)
建筑高度：106 m (348 ft)
主要功能：办公
开发商：深圳泰然（集团）股份有限公司
规划设计：筑博建筑设计集团有限公司
建筑设计：筑博建筑设计集团有限公司
景观设计：筑博建筑设计集团有限公司，Hill 景观设计公司
结构设计：筑博建筑设计集团有限公司
机电设计：筑博建筑设计集团有限公司；中国建筑科学研究院
总承包商：深圳泰然（集团）股份有限公司

frames a crystalline volume, which serves as a light-filled central entrance to the underground parking. Rather than arriving into a dark parking garage and entering the building through an unassuming elevator core, the pool provides direct access through the interior courtyard, again modifying the typical experience of arriving and entering the office environment.

A typical office tower's roof has no relationship with the public realm or the tower occupants. Taking advantage of its sloping condition, the roof of the Shenye TaiRan Building is a series of terraces, accessible on multiple levels. These spaces form the "fifth facade" of the building and are designed as a uniform landscape, despite the fact that they are all privately owned. The roof is a lush garden landscape of curving benches, wood terraces and planting, creating a unique sight from the surrounding office towers, as well as from the interior.

The vertical facade of the building is composed of stacked, elongated one floor-high units, which play a double function. First, the division enables the observer to consider each office unit in singularity, by providing option to step the glazing back, thus creating additional balcony space - an unusual feature in a typical office layout. This "in-between space" also acts as a shading device, minimizing the cooling load. Ultimately, it makes room for, and resolves the inevitable challenge of placing individual air conditioning units, without interrupting continuity of the main facade, or the roof.

This duality is further expressed in the choice of materials. The outer layer, composed of light stone, is in focus, catching and reflecting the rays of the sun. The inner balcony facade is lined with dark grey aluminum panels and remains in shadow, hiding the "contents" within.

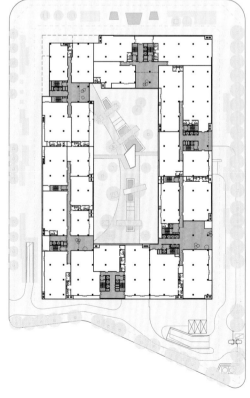

谐一致的景观。屋顶上有着繁茂的花园景观、弯曲的长椅、木质的平台和植被，无论从周围的塔楼还是从内部都能看到屋顶花园独特的景观——尽管实际上它是私有空间。

建筑物垂直的立面由一层楼高的细长单元堆叠而成，收到一举两得之效。该布局创造出相对独立的办公空间，同时也提供将玻璃幕墙移后的选择，以构建出额外的阳台。这在传统的办公室布局中并不常见。这个"间隔空间"也可以作为遮阳设备，最大限度地降低冷却负荷。总而言之，这样的设计节省了空间，还解决了一些难以避免的问题，如因放置空调而截断墙面或屋顶的情况。

这种双重考虑在材料的选择上也有进一步表现。外层墙面采用浅色石材质，旨在最大反射阳光。内阳台的立面则嵌有位于背阴面的深灰色铝板，可以有效地储蓄光能。

## Jury Statement | 评委会评语

Where so many projects place towers in a "park," at the Shenye TaiRan Building, a park climbs around the perimeter, breaking down the scale of a full-block development and providing green space at height to its occupants. This is highly suggestive of what future vertical cities could look like.

很多项目都把大厦建在"公园"里，深业泰然大厦则是在建筑周边开发了一个公园。这样做可以让这个庞大的建筑物看起来更加柔和，同时也为住户提供一抹绿色的景观。将来城市的垂直开发可以以此作为范本。

**Previous Spread | 前页**
Left: Overall view from west
左图：西面全景

**Current Spread | 本页**
Opposite Left: Detailed view of stepped garden spaces
对侧左图：阶梯花园
Opposite Right: Ground-floor plan, showing green and public spaces
对侧右图：地面层平面图，显示出公共空间与绿化
Right: Central communal space
右图：中央公共空间

*"The Shenye Tairan Building is an effective fusion between iconic architecture and sensitive urban design, offering an attractive pedestrian environment, while the terraced design adds character to the workplace."*

"深业泰然大厦有效融合了地标性建筑和城市敏感区的设计，提供了极其有趣的人行环境，同时阶梯设计又给办公区增加了亮点。"

Design Juror　设计评委

# China Urban Habitat Award – Nominee
中国高层建筑城市人居奖提名作品

## Huizhou Central Place | 惠州华贸中心
Huizhou | 惠州

The Huizhou Central Place includes a mall, three high-rise office buildings (120–188 meters), four high-rise residential buildings (130–150 meters) and a kindergarten. Because of the high density of tall buildings in the center, it was essential that public space be prioritized. The resulting central plaza has the special qualities of both private and public space, in which the connecting pedestrian path is enclosed and formed by nearby buildings. The central plaza connects to the broader community outside the complex by way of a continuous commercial interface on the ground floor, which benefits the project itself, while providing more possibilities for public activities.

In each of five groups, the buildings interlace with each other to enclose their own landscape space, each of which meets localized needs. Meanwhile, each independent landscape space within the group can be accessed from the central plaza via pedestrian paths. Auto circulation on site is limited to the perimeter, to support the vitality of the central plaza.

**Completion Date:** January 2013
**Total Land Area:** 135,276 sq m (1,456,099 sq ft)
**Total Building Footprint:** 39,484 sq m (425,002 sq ft)
**Total Open Area:** 95,792 sq m (1,031,097 sq ft)
**Total Building Area:** 716,642 sq m (7,713,870 sq ft)
**Primary Function:** Huamao Towers 1, 2, 3: office; Platinum Mansions 3,4,8 & YOHO Cloud Club 7: residential
**Developer:** Huizhou City Run and the Real Estate Development Co., Ltd.
**Urban Planner:** Perkins Eastman
**Architect:** Huamao Towers 1, 2, 3: ECADI; Platinum Mansions 3,4,8 & YOHO Cloud Club 7: The Architectural Design & Research Institute of Guangdong Province
**Landscape Architects:** Huamao Towers 1, 2, 3: ECADI; Platinum Mansions 3,4,8 & YOHO Cloud Club 7: The Architectural Design & Research Institute of Guangdong Province
**Structural Engineer:** ECADI
**MEP Engineers:** ECADI
**Main Contractor:** China Construction Third Engineering Bureau Co., Ltd.

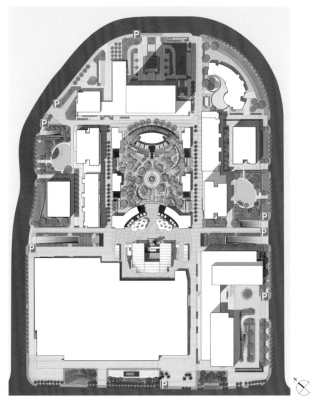

惠州华茂中心项目中，主要单体建筑（办公、住宅）均超过100 m，一方面通过超高层建筑的集群效应从整体形象上体现出现代CBD的外在气质；另一方面也在满足高容积率的要求下，尽可能地减少建筑占地面积，留出更多的底层公共空间。

在空间布局中，商业、办公楼与住宅各自形成了5个独立的组团。通过办公与住宅的高层塔楼形成一个U字形的空间界面，并与前面的商业建筑共同围合出一个中央花园广场，而这个广场通过一条贯穿整个基地的景观休闲步道与城市空间相联系，从而在整个基地内部建立起一个既公共又相对私密的市民活动空间。

竣工时间：2016年1月
总用地面积：135 276 m² (1 456 099 ft²)
总建筑基底面积：39 484 m² (425 002 ft²)
总户外面积：95 792 m² (1 031 097 ft²)
总建筑面积：716 642 m² (7 713 870 ft²)
主要功能：办公 / 居住 / 商业
开发商：惠州市华贸兴业房地产有限责任公司
规划设计：美国Perkins Eastman建筑设计事务所
建筑设计 / 景观：华东建筑设计研究总院；广东省建筑设计研究院
结构设计：华东建筑设计研究总院；广东省建筑设计研究院
机电设计：华东建筑设计研究总院；广东省建筑设计研究院
总承包商：中国建筑第三工程局有限责任公司

**Opposite Top:** Overall view of complex
对侧上图：综合体全景

**Opposite Bottom:** Exterior View
对侧下图：室外远景

**Top Left:** Building entrance
上左图：大厦入口

**Top right:** Aerial view
上右图：鸟瞰图

**Bottom Right:** Site plan
下右图：总平面图

# China Tall Building Urban Habitat Award – Nominee
中国高层建筑城市人居奖提名作品

## Wuhan Tiandi Site A | 武汉天地A座
Wuhan | 武汉

Wuhan Tiandi is an urban regeneration project that transformed a dilapidated, under-utilized district into Wuhan's premier community. The project is accessible by 34 bus lines, one existing and one planned metro line. The development adaptively re-uses historically and culturally significant buildings within its perimeter.

Unusual for a tall-building development in China, the project integrates buildings of different heights to create spatial and visual interest, to preserve views to the Yangtze River, and to create valuable settings for landmark towers. The design offers courtyard housing and small, walkable blocks, minimizes buildings' setbacks to enhance the pedestrian experience, and reserves open spaces for public use. Resolving some long-standing urban development challenges in China, Wuhan Tiandi has stimulated the economic growth and community transformation of its neighborhood, and is now a new meeting point and a natural place for public events, cultural activities and celebrations for the entire city. It is much-emulated by other developers throughout China.

**Completion Date:** December 2013
**Total Land Area:** 194,470 sq m (2,093,258 sq ft)
**Total Building Footprint:** 73,110 sq m (786,949 sq ft)
**Total Open Area:** 121,360 sq m (1,306,308 sq ft)
**Total Hardscape:** 146,340 sq m (1,575,191 sq ft)
**Total Softscape:** 48,130 sq m (518,067 sq ft)
**Developer:** Shui On Land Limited
**Urban Planner:** Skidmore, Owings & Merrill LLP
**Architect:** BenWood Studio Shanghai; P & T Group
**Landscape Architects:** Design Land Collaborative; WAA
**Structural Engineer:** P & T Group
**MEP Engineer:** P & T Group
**Main Contractor:** China Construction Third Engineering Bureau Co., Ltd.

武汉天地是一个旧城改造项目，目的在于将危房遍布的欠开发区建为武汉首屈一指的社区。作为 LEED-ND 金奖二级项目，此地有 34 条公交线路和一个轻轨站，未来还规划了地铁。该项目将其范围内历史久远、文化丰富的建筑物因地制宜地加以改造。

不同于国内一般的高层建筑，该项目内建筑高度各异，创造出了有趣的立体视觉效果，保留了长江的视野景观，还成为地标性建筑的标杆。项目设计提供了庭院式住宅和适于步行的小型街区，最大限度地减少了建筑物的阻挡，优化了行人穿越街道的体验，并预留出公共开放空间。武汉天地解决了我国一些城市发展长期存在的通病，刺激了周边的经济增长和社区转型，现在已经成为武汉市民聚会、公共活动和文化庆祝的首选之地。

竣工时间：2013 年 12 月
总用地面积：194 470 m² (2 093 258 ft²)
总建筑基底面积：73 110 m² (786 949 ft²)
总户外面积：121 360 m² (1 306 308 ft²)
总硬质景观面积：146 340 m² (1 575 191 ft²)
总软质景观面积：48 130 m² (518 067 ft²)
开发商：瑞安房地产有限公司
规划设计：SOM 建筑事务所
建筑设计：本杰明·伍德建筑事务所；巴马丹拿建筑设计咨询有限公司
景观：地茂景观设计咨询（上海）有限公司；WAA
结构设计：巴马丹拿建筑设计咨询有限公司
机电设计：巴马丹拿建筑设计咨询有限公司
总承包商：中国建筑第三工程局有限公司

**Opposite Top:** Overall view (Site A)
对侧上图：全景

**Opposite Bottom:** Aerial view of green roofs and green features
对侧下图：鸟瞰绿色屋顶与花园

**Top Left:** Main commercial hub
上左图：商业街

**Top Right:** Site plan
上右图：总平面图

**Right:** Green features incorporated throughout the site
右图：整个项目都绿树成荫

# Chinese Tall Building Legacy Award
中国高层建筑成就奖

# Awards Criteria
评选标准

This award recognizes proven value and performance over the period of time from China's economic opening in 1978 to a point up to 10 years preceding the Award year. This award gives an opportunity to reflect back on buildings that have been completed and operational for at least 10 years prior to the date of the award, and to acknowledge those projects that have performed successfully long after the ribbon-cutting ceremonies have passed.

1. The project must be physically located in the Greater China region, including Hong Kong, Macau, Taiwan and Mainland China. In its inaugural year, the building must have a completion date after January 1, 1978 and before Dec. 31, 2005.

2. The buildings' performance must be evident, and may include but is not limited to: contribution to urban realm, contribution to culture / iconography, social issues, internal environment / occupant satisfaction, technical/engineering performance (structural engineering, seismic, wind, etc.), environmental performance, fire & safety, vertical transportation, etc.

3. Projects must be considered "tall" buildings. If a project is less than 100 meters, it is unlikely to qualify.

4. In its inaugural year, ten winners and ten Honorable Distinction recipients will be selected, representing a wide range of projects completed between 1978 and 2005.

该奖项是为表彰从中国 1978 年改革开放至颁奖前 10 年的范围内建成建筑的价值和表现。这一奖项将回顾那些至少 10 年前已建成的建筑，表彰它们自投入使用后长久以来的成功运营。

1. 参加首届评奖的建筑的竣工时间必须在 1978 年 1 月 1 日至 2005 年 12 月 31 日之间。

2. 建筑应具有显著且极佳的表现，其中包括但又不仅限于：对城市的贡献、对文化 / 传统形象的贡献、社会问题、内部环境 / 居住者满意度、科技 / 工程表现（结构工程、抗震、防风工程等）、环境能效、防火与安全、垂直交通及其他。

3. 参选项目必须是"高层"建筑。如果项目的建筑高度低于 100 m，将不具备参选资格。

4. 在首届评奖活动中将诞生 10 个获奖作品和 10 个荣誉奖作品，竣工年份跨度在 1978 年至 2005 年之间。

# China Tall Building Legacy Award – Winner
# 中国高层建筑成就奖

## White Swan Hotel | 白天鹅宾馆
Guangzhou | 广州

## 1983

The White Swan Hotel represents the introduction of foreign investment in China, carrying design, construction and management to a new level. The luxury hotel is located on Shamian Island, overlooking the Pearl River and facing the White Swan Pool. The hotel is reached by its own private 635-meter (2,083-foot) causeway.

  白天鹅宾馆是中国引入外资的象征，将设计、建造与管理都带上了新台阶。该豪华宾馆坐落在沙面白鹅潭，俯瞰珠江，面朝白天鹅湖。635 m（2 083 ft）长的附属堤道与其相连。

**Completion Date:** 1983
**Height:** 100 m (328 ft)
**Use:** Hotel
**Architect:** Guangzhou Design Institute
**Developer:** Fok Ying-tung; White Swan Hotel Group Co., Ltd.
**Architect:** Guangzhou Design Institute
**Main Contractor:** Guangzhou Di-Er Construction & Engineering Co., Ltd.

竣工时间：1983 年
高度：100 m（328 ft）
主要功能：酒店
业主：白天鹅酒店集团有限公司
开发商：霍英东；白天鹅酒店集团有限公司
建筑设计：广州市设计院
总承包商：广州第二建筑工程有限公司

# China Tall Building Legacy Award – Winner
中国高层建筑成就奖

## Hong Kong and Shanghai Bank | 汇丰银行
Hong Kong | 香港

## 1985

The Hong Kong and Shanghai Banking Corporation (HSBC) headquarters building is one of the world's most recognizable skyscrapers, with its extensive exterior bracing system giving the appearance of an industrial facility or a cargo gantry, perhaps a reference to its owner's deep roots in Hong Kong's development as a global commercial center. On the ground floor, the open-air pedestrian passageway, with its views up into the suspended banking hall through a glass ceiling, communicates a metaphorical and physical transparency that has endeared the building to the public.

汇丰银行（HSBC）总部大楼是世界上最具辨识度的摩天大楼之一，宽阔的外部结构支撑系统使其看起来类似工业设施或货物起重机，作为全球商业中心，它似乎映射了其业主在香港发展的根深蒂固。一楼设有露天人行道，通过玻璃天花板可以看到悬浮式的银行大厅，建筑对公众的开放隐喻着其体系的透明。

**Completion Date:** 1985
**Height:** 179 m (587 ft)
**Stories:** 43
**Area:** 99,171 sq m (1,067,468 sq ft)
**Use:** Office
**Architect:** Foster and Partners
**Structural Engineer:** Arup; Cleveland Bridge Ltd.
**MEP Engineer:** J. Roger Preston Group

竣工时间：1985 年
高度：179 m（587 ft）
层数：43
面积：99 171 m²（1 067 468 ft²）
主要功能：办公
建筑设计：福斯特建筑事务所
结构设计：奥雅纳工程咨询有限公司；克利夫兰桥梁建筑公司
机电设计：澧信工程顾问有限公司

# China Tall Building Legacy Award – Winner
中国高层建筑成就奖

## International Foreign Trade Center
国贸中心

Shenzhen | 深圳

**1985**

Located at the junction of Jiabin Road and Renmin South Road, this building was completed in 37 months, earning the city's rapid rate of development the nickname "Shenzhen Speed." It was the tallest building in China upon completion. It was also an important place to visit in Shenzhen.

该建筑位于深圳市嘉宾路与人民南路交叉口，历时37个月就竣工了，被称为"深圳速度"，意味着跟深圳的发展变化一样快。它是当时全国最高的建筑，也是深圳重要景点之一。

**Completion Date:** December 1985
**Height:** 160 m (525 ft)
**Stories:** 50
**Use:** Hotel
**Developer:** Shenzhen Property Development
**Architect:** Central-South Architectural Design Institute Co., Ltd.

竣工时间：1985年12月
高度：160 m (525 ft)
层数：50
主要功能：酒店
开发商：深房集团
建筑设计：中南建筑设计研究院

# China Tall Building Legacy Award – Winner
## 中国高层建筑成就奖

## East China Electrical Power Distribution Building
## 华东电力调度大楼

Shanghai ｜上海
### 1989

Back in 1980s when it was built, East China Electrical Power Distribution Building was a localized Chinese skyscraper with innovative elements. With "quarter window" "pitched roof" and "brown wall brick", it puts the architectural culture of Shanghai into the building design, breaking regular office building design and becoming a building of unique external form on Nanjing Road. Different from the traditional linear layout, this building, an excellent representative of Shanghai construction in the 1980s, formed an 45° angle against the street which not only frees neighborhood from its shadow but also leave a buffer space between the building and street. It successfully broke the closed and suppressive feeling of Nanjing Road at that time.

  华东电力调度大楼是20世纪80年代建造的中国本土化原创的摩天楼。"三角窗"、"斜屋顶"、"棕墙砖"的外观将上海建筑文化融入到大楼中，打破了中规中矩的办公楼设计，在南京路上独树一帜。大厦45°沿街转身，突破了传统的线性布置方式，在不遮挡附近街区阳光的同时，让建筑与街面留有缓冲空间，打破了当时南京路封闭压抑的感觉，是80年代上海建筑的杰出代表。

**Completion Date:** 1989
**Height:** 126 m (412 ft)
**Stories:** 26
**Use:** Office
**Owner/Developer:** State Grid Corporation of China
**Architect:** Elsewhere we refer to it as ECADI. We should be consistent in this.

竣工时间：1989 年
高度：126 m（412 ft）
层数：26
主要功能：办公
业主/开发商：国家电网
建筑设计：华东建筑设计研究总院

# China Tall Building Legacy Award – Winner
## 中国高层建筑成就奖

## Bank of China Tower | 中国银行大厦
Hong Kong | 香港

## 1990

One of Hong Kong's most distinctive buildings, Bank of China Tower is situated in the center of the city's skyline. At the ground level, the tower is set back from the street to create an inviting pedestrian environment that is convenient, yet secluded from urban congestion. It is surrounded by a broad promenade, and flanked by cooling water gardens. The building is known for its observation deck on the 43rd floor.

作为香港最与众不同的建筑之一，中国银行大厦坐落于城市轮廓的中央。该建筑的一楼设置在街道后方，在拥挤的城区中创造出了有趣、便利又隐蔽的人行道。周围环绕着一条宽广的散步区，侧翼设有水上花园。建筑内第43层的观景台远近闻名。

**Completion Date:** May 1990
**Height:** 367 m (1,205 ft)
**Stories:** 72
**Area:** 135,000 sq m (1,453,128 sq ft)
**Use:** Office
**Owner:** Bank of China
**Developer:** Bank of China
**Architect:** I.M. Pei & Partners; Kung & Lee Architects Ltd.
**Structural Engineer:** Leslie E. Robertson Associates; Valentine, Laurie, and Davis
**MEP Engineer:** Jaros, Baum & Bolles; Associated Consulting Engineers
**Main Contractor:** Hong Kong Construction (Holdings) Limited; Kumagai Gumi

竣工时间：1990年5月
高度：367 m（1 205 ft）
层数：72
面积：135 000 m²（1 453 128 ft²）
主要功能：办公
业主/开发商：中国银行
建筑设计：贝聿铭建筑设计事务所；Kung & Lee 建筑设计有限公司
结构设计：理雅结构工程咨询有限公司；Valentine, Laurie, and Davis
机电设计：JB&B 工程师事务所；Associated 工程顾问公司
总承包商：香港建设（控股）有限公司；日本株式会社熊谷组

# China Tall Building Legacy Award – Winner
中国高层建筑成就奖

## Shanghai Center | 上海商城
Shanghai | 上海

## 1990

Opposite to Shanghai Exhibition Center, Shanghai Center is located in Jing'an shopping district. It is one of the earliest high-rise complexes in Shanghai, and now serves as a hotel, a service apartment, an office, a commercial area and performing stage. It has an exquisite well-proportioned modeling image, and simple but elegant painted facade. In addition, the Chinese garden in its courtyard, columns and arch on its first floor, and winding corridors and railing on second floor all show a respect to the local architectural culture.

上海商城位于上海市静安购物商业区，上海展览中心对面，是上海最早建成的超高层建筑综合体之一，该项目包含了酒店、服务式公寓、办公、商业、观演等功能。造型比例精当，涂料装饰的立面朴素而不失优雅。院落里中国园林的引入，首层的立柱、拱门，二楼商业的曲折回廊和栏杆，都表达了一种对本地建筑文化的尊重。

**Completion Date:** 1990
**Height:** Portman Ritz-Carlton: 165 m (541 ft); Shanghai Center East and West Apartments: 112 m (367 ft)
**Stories:** Portman Ritz-Carlton: 48; Shanghai Center East and West Apartments: 34
**Use:** Portman Ritz-Carlton: Hotel; Shanghai Center East and West Apartments: Serviced apartments
**Architect:** John Portman & Associates (design); ECADI (architect of record)
**Structural Engineer:** John Portman & Associates

竣工时间：1990 年
高度：波特曼丽嘉酒店：165 m (541 ft)；上海商城东部和西部公寓：112 m (367 ft)
层数：波特曼丽嘉酒店：48；上海商城东部和西部公寓：34
主要功能：波特曼丽嘉酒店：酒店；上海商城东部和西部公寓：酒店式公寓
建筑设计：约翰·波特曼建筑设计事务所；华东建筑设计研究总院

# China Tall Building Legacy Award – Winner
中国高层建筑成就奖

## Shun Hing Square | 信兴广场
Shenzhen | 深圳

### 1996

Shun Hing Square is a key feature of Shenzhen's Special Economic Zone. The tower and its nearby annex are juxtaposed in a "T" formation to accommodate mutual sightlines. The main tower was built at an incredible rate of four floors every nine days, taking just 40 months to complete. The design also strove to encompass the maximum fire compartment size per floor, yielding a structure with a 1:8 height-to-width ratio, an extraordinary feat in this typhoon and earthquake design zone.

信兴广场是深圳经济特区的地标性建筑。该塔楼及其附近建筑呈T字形以防止遮挡视线。主楼施工速度极快，平均每9天建成4层楼，仅用40个月即竣工。该设计还力图使每层的防火区最大化，从而形成了高宽比1:8的结构，这对处于台风和地震区的深圳来说是一项壮举。

**Completion Date:** 1996
**Height:** 384 m (1,260 ft)
**Stories:** 69
**Area:** 280,000 sq m (3,013,895 sq ft)
**Use:** Office
**Owner:** Kumagai Gumi
**Developer:** Karbony Investment
**Architect:** American Design Associates; K.Y. Cheung Design Associates
**Structural Engineer:** AECOM; Leslie E. Robertson Associates
**MEP Engineer:** Associated Consulting Engineers
**Main Contractor:** China State Construction Engineering Corporation; Hong Kong Construction (Holdings) Limited; Kumagai Gumi

竣工时间：1996 年
高度：384 m（1 260 ft）
层数：69
面积：280 000 m²（3 013 895 ft²）
主要功能：办公
业主：熊谷组
开发商：Karbony 投资公司
建筑设计：美国设计有限公司；张国言设计事务所
结构设计：艾奕康建筑设计有限公司；理雅结构工程咨询有限公司
机电设计：Associated 工程顾问公司
总承包商：中国建筑工程总公司；香港建设（控股）有限公司；日本株式会社熊谷组

# China Tall Building Legacy Award – Winner
中国高层建筑成就奖

## Jin Mao Tower | 金茂大厦
Shanghai | 上海

## 1999

The Jin Mao Tower, a mixed-use complex containing offices, convention space and a hotel, in 2013 became the tallest and the longest-operated building in China to receive a LEED-EB: OM (Existing Buildings: Operations + Management) Gold certification. Its high performance has been achieved with the assistance of a computerized energy management system, which has been in place since the building opened, and is integrated with the broader enterprise asset management (EAM) system.

　　金茂大厦是一座包含办公、会议和酒店等多功能的建筑体，2013年被评为中国最高的长期经营建筑，并获得了LEED-EB：OM（既有建筑：运营＋管理）金奖。通过计算机控制的能源管理系统，其高性能运营得到了实现。该系统在大楼竣工之初就被安置好，并集成于企业资产管理（EAM）系统。

**Completion Date:** 1999
**Height:** 421 m (1,380 ft)
**Stories:** 88
**Area:** 289,500 sq m (3,116,152 sq ft)
**Use:** Hotel / Office
**Owner/Developer:** China Jin Mao Group Co., Ltd.
**Architect:** Skidmore, Owings & Merrill LLP; Institute of Shanghai Architectural Design & Research
**Structural Engineer:** Skidmore, Owings & Merrill LLP; Institute of Shanghai Architectural Design & Research
**MEP Engineer:** Skidmore, Owings & Merrill LLP
**Main Contractor:** Shanghai Construction Group

竣工时间：1999年
高度：421 m（1 380 ft）
层数：88
面积：289 500 m²（3 116 152 ft²）
主要功能：酒店／办公
业主／开发商：中国金茂（集团）股份有限公司
建筑设计：SOM建筑事务所；上海建筑设计研究院有限公司
结构设计：SOM建筑事务所；上海建筑设计研究院有限公司
机电设计：SOM建筑事务所
总承包商：上海建工集团

# China Tall Building Legacy Award – Winner
## 中国高层建筑成就奖

# Two International Finance Center
## 国际金融中心二期
Hong Kong | 香港
2003

The International Finance Center is an integrated commercial development on the waterfront of Hong Kong's Central District. The tower stands apart from the cluster of other high-rise buildings, distinguished by its unique configuration as well as its proximity to the water. Its obelisk form tapers with subtle setbacks, which evokes a sense of ascension. The scrupulous articulation of the curtain wall softens the surface and emphasizes the verticality of the form.

国际金融中心二期位于香港中环区，是一个滨水的集成式建筑。该塔楼以其独特的结构及临水距离从其他高层建筑群中脱颖而出。方尖碑的形式配以轻微的锥度后移，产生了一种上升感。幕墙的无缝衔接柔化了建筑外观，突出了其形式的立体感。

**Completion Date:** May 2003
**Height:** 412 m (1,352 ft)
**Stories:** 88
**Area:** 185,805 sq m (1,999,988 sq ft)
**Use:** Office
**Developer:** Henderson Land Development; Sun Hung Kai Properties Limited; The Hong Kong and China Gas Company Limited; MTR Corporation Limited
**Architect:** Cesar Pelli & Associates; Rocco Design Architects Limited
**Structural Engineer:** Arup (design); Leslie E. Robertson Associates (peer review)
**MEP Engineer:** J. Roger Preston Group
**Main Contractor:** E Man-Sanfield JV Construction

竣工时间：2003 年 5 月
高度：412 m（1 352 ft）
层数：88
总面积：185 805 m² （1 999 988 ft²）
主要功能：办公
开发商：恒基兆业地产有限公司；新鸿基地产集团有限公司；香港中华煤气有限公司；香港铁路有限公司
建筑设计：西萨·佩里建筑设计事务所；许李严建筑师事务有限公司
结构设计：奥雅纳工程咨询有限公司（设计）；理雅结构工程咨询有限公司（同行评审）
机电设计：澧信工程顾问有限公司
总承包商：E Man-Sanfield JV Construction

# China Tall Building Legacy Award – Winner
中国高层建筑成就奖

## TAIPEI 101 | 台北101大楼
Taipei | 台北

## 2004

TAIPEI 101 represents a worldwide precedent for sustainable skyscraper development. It achieved a LEED Platinum certification for Operations and Maintenance in 2011, an impressive feat for a tower of its size and complexity. The tower rises from its base in a series of eight-story modules that flare outward, evoking the form of a Chinese pagoda. The building was the world's tallest from 2004 to 2010.

台北 101 大楼是全球可持续发展摩天大楼的先例。它于 2011 年获得了运营和维护的 LEED 白金认证，其规模和复杂性令人叹为观止。该楼的基础是一系列向外张开的八层小楼，类似于中国宝塔的形式。台北 101 大楼是 2004 至 2010 年间的世界第一高楼。

**Completion Date:** December 2004
**Height:** 508 m (1,667 ft)
**Stories:** 101
**Area:** 198,347 sq m (2,134,989 sq ft)
**Use:** Office
**Owner/Developer:** Taipei Financial Center Corporation
**Architect:** C.Y. Lee & Partners Architects/Planners
**Structural Engineer:** Evergreen Consulting Engineering (design); Thornton Tomasetti (peer review)
**MEP Engineer:** Continental Engineering Consultants, Inc. (design); Lehr Associates (peer review)
**Main Contractor:** Kumagai Gumi; RSEA Engineering; Samsung C&T Corporation; Ta-You-Wei Construction; Taiwan Kumagai

竣工时间：2004 年 12 月
高度：508 m（1 667 ft）
层数：101
总面积：198 347 m² （2 134 989 ft²）
主要功能：办公
业主 / 开发商：台北金融中心公司
建筑设计：李祖原联合建筑师事务所
结构设计：永竣工程顾问股份有限公司（设计）；宋腾添玛沙帝工程顾问公司（同行评审）
机电设计：Continental 工程顾问有限公司（设计）；Lehr 顾问公司（同行评审）
总承包商：日本株式会社熊谷组；荣工工程股份有限公司；三星 C&T 公司；大有为营造股份有限公司；华熊营造股份有限公司

# China Tall Building Legacy Award – Honorable Distinction
## 中国高层建筑成就荣誉奖

## China Resources Building
### 华润大厦

Hong Kong | 香港

**Completion Date:** 1983
**Height:** 178 m (584 ft)
**Stories:** 50
**Area:** 99,000 sq m (1,065,627 sq ft)
**Use:** Office
**Owner/Developer:** China Resources (Holdings) Company Limited
**Architect:** WMKY Limited (design); Ronald Lu & Partners (renovation)
**Structural Engineer:** Siu Yin-Wai & Associates
**MEP Engineer:** Talent Mechanical and Electrical Engineers Ltd.; Junefair Group (retrofit design)
**Main Contractor:** CR Construction

竣工时间：1983 年
高度：178 m（584 ft）
层数：50
面积：99 000 m²（1 065 627 ft²）
主要功能：办公
业主／开发商：华润（集团）有限公司
建筑设计：云麦郭杨建筑师工程师事务所（设计）；吕元祥建筑师事务所（改造）
结构设计：邵贤伟建筑工程师事务所
机电设计：汇智机电顾问有限公司；增辉集团（改造设计）
总承包商：华润建筑有限公司

## Kunlun Hotel
### 昆仑饭店

Beijing | 北京

**Completion Date:** 1986
**Height:** 102 m (335 ft)
**Stories:** 28
**Use:** Hotel

竣工时间：1986 年
高度：102 m（335 ft）
层数：28
主要功能：酒店

# China Tall Building Legacy Award – Honorable Distinction
## 中国高层建筑成就荣誉奖

**1988**

## Lippo Center
## 力宝中心

Hong Kong | 香港

**Completion Date:** 1988
**Height:** 186 m (610 ft)
**Stories:** 48
**Use:** Office
**Architect:** Paul Rudolph; Wong & Ouyang
**Main Contractor:** Hip Hing Construction

**1988**

## Jin Jiang Tower Hotel
## 上海新锦江大酒店

Shanghai | 上海

**Completion Date:** 1988
**Height:** 153 m (502 ft)
**Stories:** 46
**Area:** 57,330 sq m (617,095 sq ft)
**Use:** Hotel
**Architect:** Wong Tung & Partners

竣工时间：1988 年
高度：186 m（610 ft）
层数：48
主要功能：办公
建筑设计：保罗·鲁道夫建筑事务所；王欧阳（香港）有限公司
总承包商：协兴建筑有限公司

竣工时间：1988 年
高度：153 m（502 ft）
层数：46
面积：57 330 m²（617 095 ft²）
主要功能：酒店
建筑设计：王董建筑师事务有限公司

# China Tall Building Legacy Award – Honorable Distinction
## 中国高层建筑成就荣誉奖

**1990**

## Jing Guang Center
## 京广中心
Beijing | 北京

**Completion Date:** 1990
**Height:** 208 m (682 ft)
**Stories:** 51
**Area:** 145,407 sq m (1,565,148 sq ft)
**Use:** Residential / Hotel / Office
**Developer:** New World Development Company Limited
**Architect:** Nihon Sekkei
**Structural Engineer:** AECOM
**MEP Engineer:** Meinhardt
**Main Contractor:** Hong Kong Construction (Holdings) Limited

竣工时间：1990 年
高度：208 m（682 ft）
层数：51
面积：145 407 m²（1 565 148 ft²）
主要功能：住宅 / 酒店 / 办公
开发商：新世界发展有限公司
建筑设计：株式会社日本设计
结构设计：艾奕康建筑设计有限公司
机电设计：迈进工程设计咨询有限公司
总承包商：香港建设（控股）有限公司

**1992**

## Central Plaza
## 香港中环广场
Hong Kong | 香港

**Completion Date:** 1992
**Height:** 374 m (1,227 ft)
**Stories:** 78
**Area:** 130,140 sq m (1,400,815 sq ft)
**Use:** Office
**Developer:** Sino Land Company Limited; Sun Hung Kai Properties Limited
**Architect:** Dennis Lau & Ng Chun Man & Associates, Architects & Engineers China Ltd.
**Structural Engineer:** Arup
**Main Contractor:** Manloze Ltd.

竣工时间：1992 年
高度：374 m（1,227 ft）
层数：78
面积：130 140 m²（1 400 815 ft²）
主要功能：办公
开发商：信和置业有限公司；新鸿基地产发展有限公司
建筑设计：刘荣广伍振民建筑师事务所
结构设计：奥雅纳工程咨询有限公司
总承包商：Manloze 有限公司

# China Tall Building Legacy Award – Honorable Distinction
## 中国高层建筑成就荣誉奖

**1997**

## T & C Tower
### 高雄85大楼

Kaohsiung | 高雄

**Completion Date:** 1997
**Height:** 348 m (1,140 ft)
**Stories:** 85
**Area:** 306,337 sq m (3,297,384 sq ft)
**Use:** Hotel / Office / Retail
**Owner/Developer:** Chien Tai Cement Corporation, Tuntex Group
**Architect:** C.Y. Lee & Partners Architects/Planners; HOK
**Structural Engineer:** Evergreen Consulting Engineering; TY Lin International
**MEP Engineer:** Continental Engineering Consultants, Inc.; William Tao & Associates, Inc.
**Main Contractor:** Turner International LLC

竣工时间：1997 年
高度：348 m（1 140 ft）
层数：85
面积：306 337 m²（3 297 384 ft²）
主要功能：酒店 / 办公 / 商业零售
业主 / 开发商：建台水泥公司，东帝士集团
建筑设计：李祖原联合建筑师事务所；HOK 建筑师事务所
结构设计：永峻工程顾问股份有限公司；林同棪国际工程咨询有限公司
机电设计：Continental 工程顾问公司；William Tao 工程顾问公司
总承包商：Turner 国际建设公司

**2000**

## Bank of China Tower
### 上海中银大厦

Shanghai | 上海

**Completion Date:** 2000
**Height:** 226 m (742 ft)
**Stories:** 53
**Area:** 116,945 sq m (1,258,786 sq ft)
**Use:** Office
**Architect:** Nikken Sekkei Ltd.; Institute of Shanghai Architectural Design & Research
**Structural Engineer:** Nikken Sekkei Ltd.
**MEP Engineer:** Nikken Sekkei Ltd.
**Main Contractor:** China Construction Third Engineering Bureau Co., Ltd.

竣工时间：2000 年
高度：226 m（742 ft）
层数：53
面积：116 945 m²（1 258 786 ft²）
主要功能：办公
建筑设计：株式会社日建设计；上海建筑设计研究院有限公司
结构设计：株式会社日建设计
机电设计：株式会社日建设计
总承包商：中国建筑第三工程局有限公司

# China Tall Building Legacy Award – Honorable Distinction
## 中国高层建筑成就荣誉奖

## Tomorrow Square
### 明天广场
Shanghai | 上海

**Completion Date:** 2003
**Height:** 285 m (934 ft)
**Stories:** 60
**Area:** 93,000 sq m (1,001,044 sq ft)
**Use:** Residential / Hotel / Office
**Owner:** Shanghai Tomorrow Square Co., Ltd.
**Developer:** Shanghai Anlian Investment & Development Co.
**Architect:** John Portman & Associates; Institute of Shanghai Architectural Design & Research
**Structural Engineer:** John Portman & Associates; Weidlinger Associates; Shanghai Institute of Architectural Design & Research
**MEP Engineer:** Newcomb & Boyd
**Main Contractor:** Shanghai Construction No.2 (Group) Co., Ltd.

竣工时间：2003 年
高度：285 m（934 ft）
层数：60
面积：93 000 m²（1 001 044 ft²）
主要功能：住宅 / 酒店 / 办公
业主：上海明天广场有限公司
开发商：上海安联投资发展有限公司
建筑设计：约翰·波特曼建筑设计事务所；上海建筑设计研究院有限公司
结构设计：约翰·波特曼建筑设计事务所；威德林格工程师事务所；上海建筑设计研究院有限公司
机电设计：Newcomb & Boyd 工程顾问公司
总承包商：上海建工二建集团有限公司

## Guangzhou Development Center Building
### 广州发展中心大厦
Guangzhou | 广州

**Completion Date:** 2004
**Height:** 150 m (492 ft)
**Stories:** 37
**Area:** 78,600 sq m (846,043 sq ft)
**Use:** Office
**Architect:** von Gerkan, Marg and Partners Architects
**Structural Engineer:** Arup

竣工时间：2004 年
高度：150 m（492 ft）
层数：37
面积：78 600 m²（846,043 ft²）
主要功能：办公
建筑设计：GMP 建筑师事务所
结构设计：奥雅纳工程咨询有限公司

# China Tall Building Legacy Award – Nominee | 中国高层建筑成就奖提名作品

## The Landmark Gloucester Tower
## 告罗士打大厦

Hong Kong | 香港

**Completion Date:** 1980
**Height:** 160 m (523 ft)
**Stories:** 46
**Use:** Office
**Developer:** Hongkong Land Limited
**Architect:** P & T Group
**MEP Engineer:** WSP Hong Kong Ltd.

建筑高度：160 m（523 ft）
层数：46
主要功能：办公
开发商：香港置地
建筑设计：巴马丹拿建筑设计咨询有限公司
机电设计：科进香港有限公司

## Hopewell Center
## 合和中心

Hong Kong | 香港

**Completion Date:** 1981
**Height:** 222 m (728 ft)
**Stories:** 64
**Area:** 111,000 sq m (1,194,794 sq ft)
**Use:** Office
**Architect:** WMKY Limited
**Structural Engineer:** Arup

竣工时间：1981 年
高度：222 m（728 ft）
层数：64
面积：111 000 m² （1 194 794 ft²）
主要功能：办公
建筑设计：云麦郭杨建筑师工程师事务所
结构设计：奥雅纳工程咨询有限公司

## Shenzhen Development Bank
## 深圳发展银行大厦

Shenzhen | 深圳

**Completion Date:** 1996
**Height:** 184 m (602 ft)
**Stories:** 35
**Area:** 71,000 sq m (764,238 sq ft)
**Use:** Office
**Architect:** Peddle Thorp & Walker

竣工时间：1996 年
高度：184 m（602 ft）
层数：35
面积：71 000 m² （764 238 ft²）
主要功能：办公
建筑设计：柏涛建筑设计集团

# China Tall Building Legacy Award – Nominee | 中国高层建筑成就奖提名作品

## CITIC Plaza
### 广州中信广场

Guangzhou | 广州

**Completion Date:** 1996
**Height:** 390 m (1,280 ft)
**Stories:** 80
**Area:** 205,239 sq m (2,209,174 sq ft)
**Use:** Office
**Developer:** China International Trust and Investment
**Architect:** Dennis Lau & Ng Chun Man Architects & Engineers (HK) Ltd. (DLN)
**Structural Engineer:** Maunsell AECOM Group
**MEP Engineer:** Gammon Construction Limited; Tridant Engineering Company Limited
**Main Contractor:** Hong Kong Construction (Holdings) Limited; Kumagai Gammon Joint Venture

竣工时间：1996 年
高度：390 m（1 280 ft）
层数：80
面积：205 239 m² (2 209 174 ft²)
主要功能：办公
开发商：中国国际信托投资有限公司
建筑：刘荣广伍振民建筑师事务所
结构设计：茂盛 AECOM 集团
机电设计：金门建筑有限公司；Tridant 工程有限公司
总承包商：香港建设（控股）有限公司；熊谷金门合资公司

## The Center
### 香港中环中心

Hong Kong | 香港

**Completion Date:** 1998
**Height:** 346 m (1,135 ft)
**Stories:** 73
**Area:** 130,032 sq m (1,399,653 sq ft)
**Use:** Office
**Developer:** Cheung Kong Holdings; Land Development Corporation
**Architect:** Dennis Lau & Ng Chun Man Architects & Engineers (HK) Ltd. (DLN)
**Structural Engineer:** AECOM
**MEP Engineer:** Associated Consulting Engineers; Parsons Brinckerhoff Consultants Private Limited
**Main Contractor:** Paul Y - ITC Construction

竣工时间：1998 年
高度：346 m（1 135 ft）
层数：73
面积：130 032 m²（1 399 653 ft²)
主要功能：办公
开发商：长江实业；土地发展公司
建筑设计：刘荣广伍振民建筑师事务所
结构设计：艾奕康建筑设计有限公司
机电设计：Associated 工程顾问公司；柏诚顾问有限公司
总承包商：保华德祥建筑集团

## Dalian World Trade Center
### 大连世贸大厦

Dalian | 大连

**Completion Date:** 2000
**Height:** 242 m (794 ft)
**Stories:** 50
**Area:** 94,866 sq m (1,021,129 sq ft)
**Use:** Office
**Architect:** Nadel

竣工时间：2000 年
高度：242 m（794 ft）
层数：50
面积：94 866 m²（1 021 129 ft²)
主要功能：办公
建筑设计：美国纳德尔建筑设计公司

# China Tall Building Legacy Award – Nominee | 中国高层建筑成就奖提名作品

## Macau Tower
## 澳门旅游塔
Macau | 澳门

**Completion Date:** 2001
**Height:** 338 m (1,109 ft)
**Stories:** 61
**Use:** Telecommunications / Observation
**Owner:** Sociedade de Turismo e Diversoes de Macau
**Developer:** Sociedade de Turismo e Diversoes de Macau
**Architect:** CCM Architects
**Structural Engineer:** Beca Group
**MEP Engineer:** Beca Group

竣工时间：2001 年
高度：338 m（1 109 ft）
层数：61
主要功能：通信 / 观光
业主 / 开发商：澳门旅游娱乐股份有限公司
建筑设计：CCM 建筑事务所
结构设计：贝科工程设计集团
机电设计：贝科工程设计集团

## Pudong International Information Port
## 浦东国际信息港
Shanghai | 上海

**Completion Date:** 2001
**Height:** 211 m (692 ft)
**Stories:** 41
**Area:** 101,188 sq m (1,089,179 sq ft)
**Use:** Office
**Owner:** Shanghai Information World Co., Ltd.
**Developer:** Shanghai Information World Co., Ltd.
**Architect:** Nikken Sekkei Ltd.; Institute of Shanghai Architectural Design & Research
**Structural Engineer:** Nikken Sekkei Ltd.
**MEP Engineer:** Nikken Sekkei Ltd.
**Main Contractor:** China Construction Fourth Engineering Division Corp. Ltd.

竣工时间：2001
高度：211 m（692 ft）
层数：41
面积：101 188 m²（1 089 179 ft²）
主要功能：办公
业主 / 开发商：上海信息世界有限公司
建筑设计：株式会社日建设计；上海建筑设计研究院有限公司
结构设计：株式会社日建设计
机电设计：株式会社日建设计
总承包商：中国建筑第四工程局有限公司

## Guangming Building
## 光明大厦
Shanghai | 上海

**Completion Date:** 2003
**Height:** 150 m (492 ft)
**Stories:** 34
**Use:** Office
**Architect:** ECADI (Design); C.Y. Lee & Partners Architects/Planners (renovation)

竣工时间：2003 年
高度：150 m（492 ft）
层数：34
主要功能：办公
建筑设计：华东建筑设计研究总院（设计）；李祖原联合建筑师事务所（改造）

# China Tall Building Innovation Award
中国高层建筑创新奖

# Awards Criteria
评选标准

This award recognizes a specific area of recent innovation in a tall building project that has been incorporated into the design, or implemented during construction, operation, or refurbishment. Unlike the China Best Tall Building award, which considers each project holistically, this award is focused on one special area of innovation within the design, construction, or operation of the project – not the building overall. The areas of innovation can embrace any discipline, including but not limited to: technical breakthroughs, construction methods, design approaches, urban planning, building systems, façades, interior environment, etc. The important criteria for judging are that the submission outlines succinctly the area of innovation, in comparison to standard benchmarks.

1. The project must be physically located in the Greater China region, including Hong Kong, Macau, Taiwan and Mainland China.

2. The Innovation Award can include a project completed (topped out architecturally, fully clad, and at least partially occupied) no earlier than January 1st of the previous year, and no later than the current year's submission deadline (e.g., for the 2016 awards, a project must have a completion date between January 1, 2014 and December 14, 2015). The Innovation Award can also include recognition of a breakthrough that may not yet have been implemented in a specific building, but has been thoroughly tested.

3. Projects must be considered "tall" buildings. If a project is less than 100 meters, it is unlikely to qualify.

4. The project must clearly demonstrate a specific area of innovation within the design and/or construction that is new and pushes the design of tall buildings to a higher level. The area of innovation should demonstrate an element of adaptability that would allow it to influence future tall building design, construction or operation in a positive way.

5. This award is intended to recognize a single element of innovation within a project. If a single project has multiple innovative elements, these must be submitted as separate nominations.

6. Buildings that are submitted for the China Best Tall Building award are also eligible to have their specific innovations submitted to the Innovation Award in the same or subsequent years.

7. The applying company or institutions should submit corresponding qualification certificates issued by the Ministry of Housing and Urban-Rural Construction. Foreign investment enterprises should submit the corresponding certificates issued by the related authorities. For projects in Hong Kong, Macau and Taiwan, the applicant company or institution should submit the corresponding local certificates.

8. When a submittal is completed by several design institutions, the applying organization shall have agreed with the cooperating organizations the list of all participants to be credited in a "Project Joint Completion Statement.", The Project Joint Completion Statement joint completion report' shall be chopped by all named parties in the submittal.

该奖项是为奖励在高层建筑项目的设计、施工、运营或改建时使用的创新技术而设立的。与中国最佳高层建筑奖所不同的是，中国最佳高层建筑奖是考虑该项目的整体情况，而本奖项的特点是专注于高层建筑项目在设计、建设、运营中的某项专项技术创新情况，非建筑整体。创新的范围包括但不仅限于以下方面：技术突破，设计方法，施工方法，建筑物外观，外部城市关系，内部环境等。此奖项的重要评判标准要求申报项目应清晰阐述其创新领域与一般标准比较后的不同之处。

1. 该项目的地点应在大中华地区，包括中国大陆、香港、澳门和台湾。

2. 该项目可以是已竣工完成的项目（建筑物已经落成，外层装饰全部完成，并至少处于部分使用状态）。完成时间不早于上一年的 1 月 1 日，也不晚于当年的申请截止日（例如，参赛 2016 年度的项目，完成时间需在 2014 年 1 月 1 日和 2015 年 12 月 14 日之间）。该创新奖也可以包括在某一特定建筑中虽还未实施的、但已通过全部详细测试的技术创新。

3. 参选项目必须是"高层"建筑。如果项目的建筑高度低于 100 m，将不具备参选资格。

4. 申报项目需清晰地展示设计和 / 或施工中具体的创新内容，而且该创新应是最前沿的，并能推动高层建筑的设计向更高水平发展。同时，该创新内容应对未来的高层建筑设计、施工或运营方面产生积极影响。

5. 该奖项旨在奖励项目中某个创新元素，如果一个项目中有多个创新元素，则需分别申请。

6. 参加申报中国最佳高层建筑奖的项目也可以在同一年或随后的年份中将其特定的创新用于申报创新奖。

7. 申报单位或组织应提交住房和城乡建设部门颁发的相应的资质证书。外商投资企业应提交有关部门颁发的相应的证书。对于港澳台地区的项目，申报单位或组织应提交相应的本地证书。

8. 当所申报的项目由若干个设计单位共同完成时，申报主体单位应与各合作单位在"项目联合完成声明"中就参与设计的各单位的名单达成一致。名单所列各方需在该申报项目的"项目联合完成报告书"中盖章。

# China Tall Building Innovation Award – Winner
中国高层建筑创新奖

## Mega-Suspended Curtain Wall (Shanghai Tower, Shanghai)
悬挂式巨型玻璃幕墙（上海中心大厦）

The Shanghai Tower incorporates two independent curtain wall systems, which ultimately act together to deliver its signature design characteristic, a series of glass-walled atria located every 12 to 15 floors along the height of the building. The inner skin is circular, while the outer skin is a rounded triangular shape that twists and tapers along the vertical axis, creating the distinctive spiral shape of the facade.

The mega-suspended curtain wall support structural system consists of sag rods, ring beams and radial struts. The outer exterior curtain wall is supported by a series of ring beams that are shaped as rounded triangles in plan and twist around the inner tower. Horizontal forces are transferred by radial struts back to the inner circular floor slabs. Gravity forces are taken by sag rods aligned with the radial struts and hung from the MEP floors at the top of each zone. Except for the top suspension joints, the connection joints between the outer curtain wall and the main structure are flexible connections, which can accommodate differential deformation between the curtain wall support system (CWSS) and the main structure under various horizontal and vertical loads.

To ensure the successful construction of this project, several special cases were analyzed and researched in detail, such as the behavior of the flexible connection joint between the CWSS and the main structure, the relative deformation under horizontal and vertical loads, and the mechanical behavior of the system during the construction process.

"上海中心大厦"采用独立双层的玻璃幕墙系统，这是其标志性的设计特征。每隔12~15层有一个由玻璃墙组成的空中中庭。内表面为圆形，外表面为圆角三角形，并沿着垂直轴扭转收缩上升，塑造了上海中心大厦独特的螺旋式的外立面建筑外形。

悬挂式巨型玻璃幕墙支撑结构系统包含吊杆、环梁和水平径向支撑。外层幕墙由一系列平面形状为圆角三角形并绕内部塔楼旋转的环梁支撑。水平力通过径向支撑传递至内部的圆形楼面。幕墙重力荷载通过悬挂在每区顶部设备层楼面的吊杆承担。除了顶部的悬挂节点，幕墙与主体结构的其他连接节点均采用柔性连接，以吸收幕墙支撑结构（CWSS）与主体结构之间的相对变形。

设计团队对幕墙支撑结构（CWSS）与主体结构之间的柔性连接节点，水平和竖向荷载下幕墙与主体结构之间的相对变形，以及整个结构在幕墙施工过程中的力学特性等一系列特殊问题进行了详细分析，以确保项目的成功实施。

**Innovation Design Team:**
Shanghai Tower Construction & Development Co., Ltd. (Developer)
Gensler / Tongji Architectural Design (Group) Co., Ltd. (Architect)
Tongji Architectural Design (Group) Co., Ltd. / Thornton Tomasetti (Structural Engineer)

相关企业
开发商：上海中心大厦建设发展有限公司
建筑设计：晋思；同济大学建筑设计研究院（集团）有限公司
结构设计：同济大学建筑设计研究院（集团）有限公司；宋腾添玛沙帝工程顾问公司

**Left:** The suspended curtain wall support structure is visible from the atria
左图：从中庭可见悬挂式巨型玻璃幕墙

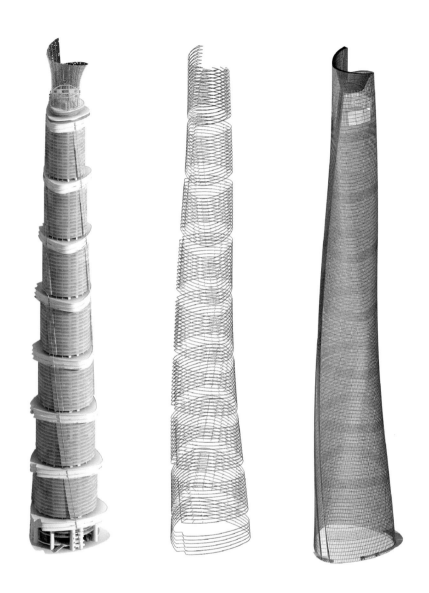

*"Rather than act as a rigid barrier between 'inside' and 'out,' this flexible solution enables the provision of 'outdoor' amenities at unprecedented heights in one of the world's tallest buildings."*

"该建筑没有严格的'内'、'外'界限,这种灵活的方式使得'户外'设施被设置在这一最高建筑中前所未有的高度上。"

Technical Juror 技术评委

This CWSS is not involved in resisting lateral load, but provides firm support for the outer curtain wall to transfer gravity and lateral loads to the main structure reliably, while being simple and light enough to minimize view obstructions. The configuration also maintains sufficient aerodynamic performance. Shanghai Tower is the first high-rise application of a suspended curtain wall structure with an aerodynamic configuration, resulting in a reduction of 25% of the wind load that would otherwise be applied to a conventional, squared-off tapering structure.

悬挂式幕墙支撑结构不参与整体结构抗侧,因而结构简洁、轻盈、视觉阻碍小,且同时可实现良好的空气动力学外形。上海中心大厦是空气动力造型悬挂式幕墙结构在超高层建筑中的首次应用,其建筑造型较常规的正方形截面的锥体造型可减少25%的风荷载。

**Right:** Overall view of the tower showing its spiral form in an urban context
左图: 上海中心呈螺旋形融入城市环境中

**Left (Opposite page):** A structural diagram of the double-skin curtain wall system
上右图: 双层幕墙结构示意图

## Jury Statement | 评委会评语

The mega-suspended curtain wall at Shanghai Tower is a formidable piece of engineering that enables the creation of peerless interior spaces and an instantly recognizable and gracious exterior shape. It does all of this while providing the aerodynamics that make the building possible in the first place. While it is a "bespoke" solution, the principles it has put into practice can be applied to the design of future tall buildings. The lessons are these: elegant solutions can be found for practical technical issues, and the quality of shared interior spaces at height is important enough to justify new heights in engineering.

上海中心大厦的巨型悬挂幕墙是一个非常艰巨的工程,它创造了一个无与伦比的室内空间和辨识度极高的优雅外观。此外,它还是这个建筑得以存在的空气动力学的一个重要组件。虽然这是一个"定制"的解决方案,然而它的实践原则可以应用到未来的高层建筑设计之中。这个设计告诉我们,解决实际技术问题的方案也可以兼顾外观的优雅。此外,共享高度空间的室内设计也体现出极高的工程技术含金量。

# China Tall Building Innovation Award – Honorable Distinction
中国高层建筑创新荣誉奖

## Cold-Bending Glass at the Supertall Scale (Nanchang Greenland Central Plaza, Parcel A)
超高层冷弯玻璃（南昌绿地中心 A 座）

The curved forms of Jiangxi Nanchang Greenland Central Plaza, Parcel A towers in Nanchang resulted from using cold-bent glass, a strategy that produces fluid three-dimensional shapes. Each glass panel that composes the façades has been warped up to 1.5% out of plane to achieve a consistent reflective, luminous appearance. The design team conducted extensive research to determine how far to push each panel out of plane in order to ensure long-term durability and architectural soundness of the panels and their sealants.

Although several other strategies were reviewed, cold-bending was the superior option, because panels produced via this process most closely approximate the smooth façade surface. Cost savings were achieved by eliminating the added material and complexity required by other options, and the geometry was simplified by allowing all four corners to exist in a single plane. The resulting curving design marks the Parcel A project as one of the earliest within architecture to apply cold-bending at the supertall scale.

**Innovation Design Team:**
Skidmore, Owings & Merrill LLP, Research, Design, and Application

创新设计团队：
SOM建筑事务所

**Left:** Workers guide panels into place during installation
左图：工人们正在安装玻璃幕墙
**Opposite Left:** A cold-bending analysis diagram shows panel warping
对侧左图：幕墙弯曲的冷弯图解
**Opposite Right:** A view of the crown structure shows the cold-bent glass in application
对侧右图：冠顶结构应用了冷弯玻璃

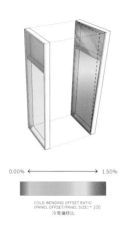
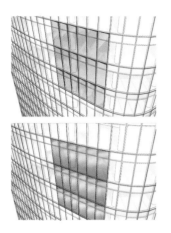

0.00% ← → 1.50%
COLD-BENDING OFFSET RATIO
(PANEL OFFSET/PANEL SIZE) * 100
冷弯偏移比

江西南昌绿地中心广场弯曲形态的奥妙源于其对冷弯玻璃的运用。这是一种能够产生尽可能流畅的三维造型的策略。每块组成外墙的玻璃板平面上向外扭曲了1.5%，从而实现了反光闪亮的整体外观。设计团队进行了大量研究来确定各块玻璃板超出平面的角度，保证了玻璃板和粘合剂的长期耐用性和结构可靠性。

尽管设计方也评估了其他几种策略，但冷弯玻璃仍然以其接近光滑表面的外观特性成为最佳选择。由于不需要像其他策略那样添加材料或复杂设计，因此成本有所降低。通过让四个角组成一个弯曲的平面，该建筑的几何平面也得到了简化。由此产生的弯曲造型标志着它成为了最早应用冷弯技术的超高层建筑项目之一。

*"The implementation of this method at an unprecedented scale means that we can now expect to see a future of smoothly engineered, aerodynamic structures that maximize efficiency and beauty."*

*"这种方法的实施是空前的，它意味着我们未来有望看到效率和美观最大化的流线型的建筑物。"*

Technical Juror　技术评委

# China Tall Building Innovation Award – Honorable Distinction
# 中国高层建筑创新荣誉奖

## Construction-Phase Internal Force Redistribution (Wuhan Center, Wuhan)
## 利用施工程序调整超高层结构内力分布的设计方法（武汉中心）

In high-rises using a mega-column structural system, when substructures are supported by trusses from above and transfer beams from below, the transfer beam might be size-restricted, due to space layout requirements and desired floor height clearances, or for the purpose of optimizing stiffness distribution. The restricted beam size would have insufficient capacity to support those loads distributed on it, whereas the capacity of the truss could be increased.

Rearranging the construction sequence can take advantage of this condition by reducing the loads distributed on the transfer beam, rather than increasing its size. The process involves using an analytical software model to reassign load paths to reduce internal forces in under-equipped transfer beams. During construction, select sub-columns are disconnected by using hydraulic jacks on specified floors after the primary structure is finished, then reconnected via welds or bolts once the load paths have been satisfactorily re-established. This technique was successfully undertaken on the 438-meter Wuhan Center, whose structure is a mega-column and frame-tube / outrigger system.

**Innovation Design Team:**
East China Architectural Design & Research Institute (ECADI)
China Construction Third Engineering Bureau Co., Ltd.

创新设计团队：
华东建筑设计研究总院
中国建筑第三工程局集团有限公司

**Left:** Overall view of Wuhan Center under construction
左图：通过施工过程控制实现结构内力重分布
**Opposite Top:** The temporary support for transfer beams during construction is visible
对侧上图：施工时可见的底部转换梁的临时支撑
**Opposite Bottom:** View of hydraulic jacks used to disconnect sub-columns
对侧下图：用于断开次结构柱的千斤顶

在包含巨型柱结构体系的超高层中，当其某一区段的次结构是由顶部的桁架与底部的转换梁共同承担时，转换梁的截面尺寸可能由于转换梁下楼层净高的要求，或基于对结构整体刚度分布合理性的考虑而受到限制。这种情况下，转换梁的承载力可能无法满足分配到其上的荷载的需求，而顶部桁架的承载能力往往有条件进行提高。

通过设定施工顺序的方法，设计人员可以主动控制传至转换梁的负荷，从而使其在截面受限制的条件下满足强度要求。该方法通过数值模拟分析确定能够减小转换梁内力的荷载传递调整路径，在施工至结构封顶后，将预先设置在指定楼层次结构柱中的受压千斤顶释放，暂时断开这些次结构柱的传力路径，使荷载在顶部桁架和底部转换梁间按设计要求实现重新分布，然后再将次结构柱焊接或栓接连接，使整体结构重新形成。该技术已在高度为438 m、采用巨型柱框筒/伸臂系统的武汉中心项目上成功实施。

*"By reducing the sizing requirement for transfer structures, this innovation paves the way for better floor-space efficiency and reduced embodied energy in fabrication."*

*"通过减少支托结构的尺寸要求，这种创新为制造业提高空间使用效率、降低能耗铺平了道路。"*

Technical Juror　技术评委

# China Tall Building Innovation Award – Honorable Distinction
中国高层建筑创新荣誉奖

## Hybrid Outrigger System (Raffles City Chongqing)
创新组合伸臂系统（重庆来福士广场）

In the design of high-rise buildings, stiffness satisfies story-drift tolerances and comfort requirements under lateral loading, while desirable seismic performance is achieved by high ductility. Conventionally, these are oppositional properties inherent to one system or another. The hybrid outrigger system effectively combines the structural stiffness advantages of a reinforced-concrete outrigger wall with the energy dissipation advantages of a fuse connection.

Components include a mild-steel shear energy dissipation element connected to the perimeter column, reinforced-concrete cantilever walls extending from the core wall, steel braces connecting the dissipation element and cantilever walls, and reinforced concrete ring beams around the core wall. The system takes advantages of the high stiffness of the concrete cantilever wall to raise the stiffness of the whole structure, while the dissipation element acts as a "fuse" during the rare earthquake, in which the ductility and energy dissipation capacity of the fuse's yield can be used to protect high-rises.

**Innovation Design Team:**
Arup

创新设计团队：
奥雅纳工程咨询（上海）有限公司

**Left:** Exterior overview of towers featuring hybrid outriggers
左图：创新组合伸臂系统使建筑外观非常有特色

**Opposite Top Left:** Steel structure of hybrid outrigger system in the laboratory
对侧上左图：组合伸臂系统的钢结构部分在实验室加工制作

**Opposite Top Right:** 3-D diagram of outrigger system
对侧上右图：组合伸臂系统的3D模型

**Opposite Bottom:** Diagram of structural components – core wall, mega-column with belt truss, second frame and hybrid outrigger
对侧下图：北部塔楼的结构元素图解——核心筒、带腰桁架的巨柱、次框架和组合伸臂

在高层建筑的设计中，针对水平荷载作用下的层间位移，需要提高刚度；而要想拥有理想的抗震性能，那就需要高延性。按照常理来说，两者截然相反，不可兼得。但是，组合伸臂系统同时采用了钢筋混凝土伸壁墙与耗能件熔断器，兼采前者高刚性与后者散能的优势，从而确保了地震中的延性，同时降低了对结构材料的要求。

组合伸臂系统包括连接于外框柱上的软钢耗能构件，从核心筒伸出的钢筋混凝土伸臂墙，连接耗能构件与伸臂墙的钢支撑，以及绕核心筒一周的钢筋混凝土环梁，既利用混凝土伸臂墙高刚度的优势提高了整体结构的刚度，耗能元件又能在大震下起到"保险丝"的作用，利用延性和耗能能力来保护高层建筑。

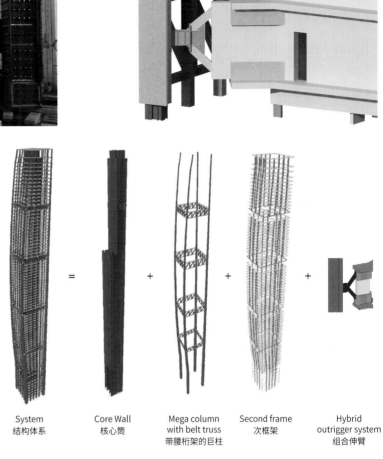

| System 结构体系 | = | Core Wall 核心筒 | + | Mega column with belt truss 带腰桁架的巨柱 | + | Second frame 次框架 | + | Hybrid outrigger system 组合伸臂 |

*"It is fitting that this safety innovation should debut in a building that appears to defy, if not the laws of gravity, then of convention. This should inspire confidence to undertake even more daring projects."*

"该创新方案在建筑中首次亮相正合时机，如果说它不是在对抗万有引力定律，那么即是打破常规。它将鼓舞更多大胆的项目付诸实施。"

Technical Juror　技术评委

# China Tall Building Innovation Award – Nominee
中国高层建筑创新奖提名作品

## Curved Façade Connector (Wuhan Center, Wuhan)
曲面建筑表皮连接构造（武汉中心）

Wuhan Center is located in Wang Jia Dun CBD area of Wuhan, China. Due to its sailing vessel image tower of 438m high beside the river, the elegant building is named as "sailing vessel city". The sailing vessel hauling upon the wind with hope and power represents thd developer marching forward courageously in the wave of economy development.

The curved façade connector developed for the Wuhan Center presents a solution to the increasingly common problem of resolving non-coplanar curves on tall buildings. The tower skin was to undulate between shallow and sharp curves, with large and small radii. This would be difficult to render effectively with standard curtain-wall panels, so the designers formulated spatial curved panels, each of which is subdivided into two vertical-plane standard panels and one horizontal-plane transition panel. The horizontal transition unit is placed between the two vertical planes and adjusted to the precise desired curvature radius. The triangulated forms were factory-produced, saving on construction time and cost.

**Innovation Design Team:**
East China Architectural Design & Research Institute Co., Ltd. (ECADI)

创新设计团队：
华东建筑设计研究总院

**Left:** Overview showing tower form resulting from non-coplanar curves
左图：非共面曲线构造下的塔楼全景

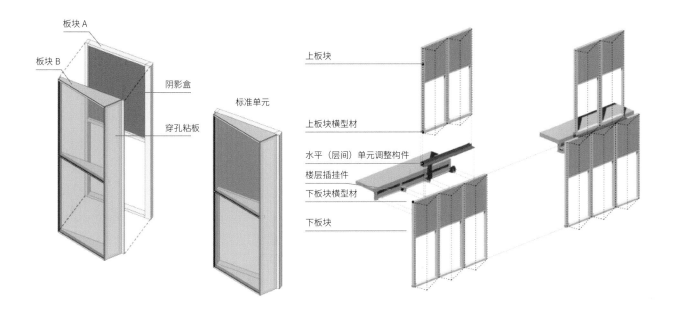

武汉中心位于武汉市王家墩CBD地区，基地临水傍湖，因高达438 m的塔楼形若帆船、造型优雅而被取名为"帆都"，寓意大厦与建设方犹如迎风张满风帆的航船，满载希望与力量，在经济建设的浪潮中乘风破浪勇往直前。

武汉中心项目在其建筑幕墙表皮的设计深化过程中开发了一种曲面建筑表皮的连接系统与构造，这一发明解决了现有传统的建筑表皮构造无法满足呈超级曲面（四点不共面）变化的建筑表皮要求。建筑师将空间形体中每一被微分的空间曲面板块（四点不共平面）解析为竖向平面标准板块和水平转换单元板块。二者相互之间的构造关系为上下竖向平面标准板块通过水平转换单元板块解决板块间竖向的空间错位与闭合的问题。每个水平转换单元板块的平面形态可被精确定义，解决了相邻板块之间各种不同的曲率半径变化的几何定义问题。同时建筑表皮单元板块能被加工成标准构件，缩短施工周期，降低建造成本，为如今日益普遍的超高层建筑非共面曲线表皮提供了良好的系统解决方案。

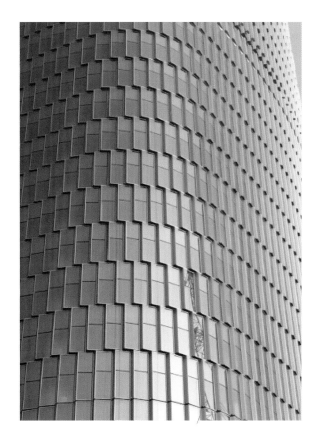

**Top:** Construction of Curtain Walls
上图：幕墙板块构造
**Bottom:** Photo of the Curtain Wall
下图：幕墙实景图

## China Tall Building Innovation Award – Nominee
## 中国高层建筑创新奖提名作品

# Gravity-Driven Fire-Extinguishing System (Zifeng Tower, Nanjing)
# 重力消防系统（南京紫峰大厦）

Well beyond the reach of ground-based firefighting equipment, a 450-meter building with a mix of uses needs alternate means of fire suppression. This system combines the transfer and reduced-pressure water tanks needed to supply water pressure, not only for firefighting, but for everyday use. During normal operation, the transfer pump brings water to the upper and lower tanks, and low-pressure water is supplied from the upper tank. In case of fire, the lower tank empties first and is replenished by gravity from the upper tank above, eliminating the need to rely on the transfer pump in case of power loss.

这座综合体建筑高达 450 m，远远超出了地面消防设备所能服务的范围，因而需要采用其他方式来承担防火的功能。该系统结合采用了传输水箱和减压水箱来供给水压，满足消防与日用需求，可谓一举两得。正常运行时，传输泵会将水输送给上下各区的水箱，低压水直接来自上水箱。遇火警时，下水箱会立即排空，随后依靠上水箱的重力重新补满，这样在断电时就不必依赖传输泵了。

**Innovation Design Team:**
East China Architectural Design & Research Institute Co., Ltd. (ECADI)

**创新设计团队：**
华东建筑设计研究总院

**Top:** Zifeng Tower waterscape feature
上图：紫峰大厦水景

**Bottom:** View highlighting height contrasts between the tower and surrounding buildings
下图：紫峰大厦与周边建筑形成高度上的极大反差

## China Tall Building Innovation Award – Nominee
## 中国高层建筑创新奖提名作品

# Hangzhou Citizen Center | 杭州市民中心

Although originally envisioned as a single 300-meter high-rise, the prevailing design of the Hangzhou Citizen Center represents a forward iteration of the "mega-city" concept as seen in Japanese Metabolism and other conceptual movements. The "mega-vertical city structure" consists of six 100-meter high-rises and four podium buildings, arranged in a circle and connected with a skybridge ring at 90 meters' height. The clustered, circular design emphasizes the centrality and importance of the project, while its permeability breaks down the scale of a superblock, and its modest height avoids overwhelming or precluding future development in the new central business district.

"杭州市民中心"最初的设想是建造一座 300 m 的超高层建筑，但其设计践行了"超级城市"这一概念，正如日本的"新陈代谢"理念及其他概念性运动所提出的。这座"巨型竖立城市结构"包括 6 座 100 m 的高层建筑及四座裙楼，排列成圆形，在 90 m 高处由环形天桥连接。紧凑的圆形设计彰显出此项工程的中心地位及重要性，但其通透性又打破了超级街区的规模，中等高度对新中央商务区未来的发展也不会造成压力和阻碍。

**Innovation Design Team:**
Tongji Architectural Design (Group) Co., Ltd. (Structural Engineer, MEP Engineer, Interior Design, Landscape Design)
Artmost Building Decoration Design Engineering Co., Ltd. (Lighting Consultant, Signage Consultant)

创新设计团队：
结构设计：同济大学建筑设计研究院（集团）有限公司
机电设计：同济大学建筑设计研究院（集团）有限公司
室内设计：同济大学建筑设计研究院（集团）有限公司
景观设计：同济大学建筑设计研究院（集团）有限公司
照明设计：Artmost 建筑装饰设计工程有限公司
标识设计：Artmost 建筑装饰设计工程有限公司

**Top:** View of the "mega-vertical city structure" with its surrounding context
上图："超级垂直城市"与其周边环境

**Bottom:** View of central roof garden and skybridges
下图：中央屋顶花园与天桥

# China Tall Building Construction Award
中国高层建筑建造奖

# Awards Criteria
评选标准

This award recognizes construction excellence in a tall building project, including but not limited to quality in construction, technical innovation, proficiency of execution of a complicated or exacting design, triumph over difficult conditions, superlative speed, efficiency, or scale. Ideally, entries would demonstrate a commendable combination of some or all of the above.

1. The project must be physically located in the Greater China region, including Hong Kong, Macau, Taiwan and Mainland China.

2. The project must be completed (topped out architecturally, fully clad, and at least partially occupied) no earlier than January 1st of the previous year, and no later than the current year's submission deadline. (i.e., for the 2016 awards, a project must have a completion date between January 1, 2014 and December 14, 2015).

3. Projects must be considered "tall" buildings. If a project is less than 100 meters, it is unlikely to qualify.

4. The project shall have no documented quality / safety problems or negative social records. Its technical complexity should be above that of its peers. The project must clearly demonstrate a specific aspects of key construction technologies, system integration, and system management that are new and push the construction of tall buildings to a higher level. Its economic, environmental and social achievements should be explicitly clear and at a high level compared to peers. The project should demonstrate an element of adaptability that would allow it to influence construction or operation of future tall buildings in a positive way.

5. Projects that are submitted for the Best Tall Building China awards are also eligible to have their specific construction aspects submitted to the Awards Program in the same or subsequent years.

该奖项是表彰高层建筑施工的杰出表现，其中包括但又不仅限于：施工质量、技术创新、对复杂与精确设计的严格执行、对困难情况的克服，或施工速度快、效率高、规模大。理想的获奖项目为以上部分或全部特点的完美结合。

1. 该项目的地点应在大中华地区，包括中国大陆和香港、澳门、台湾地区。

2. 该项目必须为已竣工完成的项目（建筑物已经落成，外层装饰全部完成，并至少处于部分使用状态）。完成时间不早于上一年的一月一日，也不晚于今年的申请截止日期（例如，参选2016年度奖项的项目，完成时间需在2014年1月1日和2015年12月14日之间）。

3. 参选项目必须是"高层"建筑。如果项目的建筑高度低于100 m，将不具备参选资格。

4. 项目建成至今没有质量和安全问题记录以及社会不良影响记录。技术复杂度超出同行水平。项目应清晰地展现在施工关键技术、系统集成和系统管理方面的一个或多个创新点，该创新点应是最前沿的，并可将高层建筑施工提升到更高的发展水平。其技术、经济、环境、社会效益指标均明显达到同行领先水平。项目应在某方面体现对未来高层建筑的施工和运营具有一定的积极作用。

5. 已经提交参加"中国最佳高层建筑"奖项评选的项目，由于在施工某些具体方面的杰出表现，也具备参加当年或随后一年的"高层建筑建造奖"的评选资格。

# China Tall Building Construction Award – Winner
中国高层建筑建造奖

## Forum 66 | 沈阳市府恒隆广场
Shenyang | 沈阳

The construction of the Forum 66 project in Shenyang combines a number of safety measures and monitoring mechanisms to overcome the challenges presented by its unique structural system. There were at least four distinctive structural and construction issues, to which the project team designed exemplary responses.

A unilaterally inclined tower design fundamentally presents some challenges in construction. The façade tilts 3.82 degrees off the vertical, resulting in the development of moment forces in several directions and locations. The project team, working with Tsinghua University, developed an analytical model based on the next-tallest leaning-tower precedent available – CCTV Headquarters in Beijing. The model not only resolved the specific problem of the Forum 66 project, but also provided a reliable reference experience for the construction of similar projects in the future.

The primary structural system is a braced-outrigger-plus-concrete frame-core tube design. The team

**Completion Date:** 2015
**Height:** 351 m (1,150 ft)
**Stories:** 68
**Owner/Developer:** Hang Lung Properties
**Project Manager:** China Construction Steel Structure Corporation Ltd.
**Main Contractor:** China Construction Third Engineering Bureau Co., Ltd.

Left: Construction of tower and retail podium
左图：塔楼与裙房正在施工中

沈阳市府恒隆广场在建造过程结合了大量的安全措施和监督机制，克服了其独特的结构体系所带来的挑战。面对该项目在结构上和施工中存在的至少四项独立的难题，项目组的攻克对策方案都具有示范性。

单面倾斜的塔楼设计为施工带来了一系列重大挑战。外立面在垂直方向上倾斜了3.82°，导致在某些位置和方向上产生了力矩。项目组和清华大学合作制作了一个分析模型，其基础是现有第二高的倾斜大厦——北京中央电视台新总部大楼。该模型不仅解决了市府恒隆广场的现有问题，也为未来相似项目的施工提供了可靠的参考经验。

工程的主结构系统采用了带伸臂桁架的劲性钢筋混凝土框架-核心筒的设计。项目组不得不解决实际施工中垂直筒使伸臂趋于非对称变形的难题。

在施工过程中，项目组采用了"延迟焊接"方案，这种方案以前在小规模项目上有过应用。通过计算机模拟、数字可视化的现场应力-应变监控与严格定时的焊接程序等多

竣工时间：2015 年
高度：351 m（1 150 ft）
层数：68
业主／开发商：恒隆地产
项目管理：中建钢构有限公司
总承包商：中国建筑第三工程局有限公司

had to counteract the tendency of vertical tubes to asymmetrically deform outriggers in practice.

During the construction process, the team implemented a "delayed welding" plan, which had been applied on smaller-scale projects previously. By using a combination of computer simulation, digital and visual on-site stress-strain monitoring, and a program of critically timed welds, the team was able to counteract the uneven-deformation tendency of the outrigger structure and frame-core tube.

Because of Shenyang's northeasterly location, the welding had to take place in temperatures of as low as -17 degrees Celsius, limiting the construction season to five months, and the finished welds needed to sustain temperatures of -30 degrees. Nearly 500 kg of welded filling needed to take place across the entire structure.

The solution was to insulate the areas being welded and carefully monitor the structural temperature while welding was taking place. By doing this, the construction season was extended, and a precedent-setting practice was established for future builders in similarly cold climates.

Quite literally, the crowning achievement of the construction process at Forum 66 was the crown structure at the top. The 53-meter-high, asymmetrical piece uses 2,000 metric tons of steel to shield nearly all the building services. The tower crane that had constructed the building up to roof level was too large to operate within the complex structure of the crown. Instead, the tower crane was used to build a smaller crane that could operate on top of the building's roof and fit within the crown's "cage" once the primary members had been erected. The smaller crane was then used to dismantle the tower crane arm, and then begin erecting the cage around itself, before being disassembled.

Through diligent application of past experience and bespoke innovation, Forum 66 rose safely above Shenyang, just as its constructors rose to the challenge of its design.

项手段，项目组得以解决伸臂结构和框架 - 核心筒的不均衡变形倾向问题。

因为地处沈阳市东北部，焊接必须在 −17 ℃ 的低温环境中进行，焊接工期仅为 5 个月，完成的焊缝需要经受 −30 ℃ 的低温。将近 500 kg 的焊接填充物需要放置在整个结构的对应位置上。

解决方案是将进行焊接的区域隔离，并在焊接过程中仔细监控结构温度。这个方案赢得了宝贵的工期时间，也为未来在同样寒冷气候里的施工提供了参考的先例。

可以肯定地说，在市府恒隆广场施工全过程中的至高成就属于建筑顶部的屋顶皇冠，53 m 高的非对称结构使用了 2 000 t 的钢材。将建筑一直建造到屋顶的大型塔吊在皇冠复杂的结构里因为体积太大而不能运作，为此，施工方用大型塔吊建造了一个适合在屋顶操作的小型塔吊。在皇冠结构的主要部件竖立起来之后，小型塔吊就能够在皇冠的"笼子"里运作。小型塔吊建成之后会先用于拆除主塔吊的吊臂，接着开始在周围建造笼状的皇冠结构。

通过对历史经验的认真研究以及量身定做的创新措施，市府恒隆广场在沈阳安然崛起，正如项目组面对设计挑战时的迎难而上一般。

**Opposite:** Construction phase – structure begins to rise above retail podium
对侧图: 商业裙房的施工阶段

**Above:** Topped-out tower during construction
上图: 主体结构封顶

> *"The four construction innovations showcased at Forum 66 solid dash when most projects, done by-the-book, would be notable for one solid dash set the bar very high for the next generation of high-rises."*

> "在大多数项目都循规蹈矩的时候，Forum 66 所展示出的四个构造创新鹤立鸡群，大大提高了下一代高层建筑的门槛。"

Technical Juror　技术评委

# China Tall Building Construction Award – Honorable Distinction
## 中国高层建筑建造荣誉奖

## Beijing Greenland Center | 北京绿地中心
Beijing | 北京

The Beijing Greenland Center contains office and serviced apartment functions, and is best known for its brocade-style shading devices. The project is commended for its creative use of a number of new technologies, such as a restrictive connection methodology for steel shear plates, an automated high-altitude spraying system, a remote quality inspection system, and its use of BIM.

The project's construction is also noted for a climbing tower crane that rises in the same core as the safety protection platform, which consolidates work activities in a smaller safety zone than in a typical building. The combination of these factors has resulted in a project that excels in both energy savings and safety.

The project designers have thus far received 11 patents for its construction methodologies, including eight patents focused on its utilities layout. Its construction techniques have earned the project the Beijing Engineering Law prize, the China Association of Construction Enterprise Management second prize, and the Huaxia Construction Science and Technology Award third prize.

**Completion Date:** Dawangjing Tower: 2015; Office Tower 1: 2016; Office Tower 2: 2016
**Height:** 260 m (853 ft)
**Stories:** 55
**Owner:** Beijing Shenglong Electric Equipment Co., Ltd.; Beijing Rundexin International Resources Investment Co., Ltd.; Zhongying Holding Group Co., Ltd.
**Developer:** Greenland Group
**Main Contractor:** Beijing NO.6 Construction Engineering Group

*"The undulating brocade façade is what will surely draw the attention of the public, but the engineering behind it is what will ultimately cement the legacy of this building."*

"起伏织锦式的外观将成为公众的焦点，而背后的工程技术则是对历史经验和过往成就的继承。"

Technical Juror　技术评委

北京绿地中心集办公楼、商业和酒店式公寓于一体。其采用单元式异型玻璃幕墙的"中国锦"自动遮阳装置是该设计的一大特点。该项目创造性地使用了一批新技术，例如钢材剪力墙约束钢板连接工艺、自动喷淋养护系统、远程质量验收系统，此外还利用了建筑信息模型技术（BIM）。

项目的建设还因与安全防护平台处于同一核心的内爬式塔吊而闻名。相较传统建筑，该塔吊将工作活动限定在了更狭窄的安全区内。这些因素的组合使得项目在节能和安全方面卓有成效。

项目设计者迄今已获得11项施工方法专利，其中8项与设施布局有关。该项目因施工技术分别获得了北京市工法审定、中国施工企业管理协会科技创新成果二等奖和华夏建设科学技术奖三等奖。

**Opposite Left:** Core tube construction using climbing formwork system
对侧左图：核心筒在建设中使用了爬模系统

**Right:** Undulating Brocade Facade
右图：起伏织锦式的外立面

竣工时间：2015/2016
高度：260 m（853 ft）
层数：55
业主：北京盛隆电气有限公司；北京润德信国际资源投资有限公司；中盈控股集团有限公司
开发商：北京绿地京华置业有限公司
总承包商：北京六建集团有限责任公司

# China Tall Building Construction Award – Nominee
中国高层建筑建造奖提名作品

## Changsha Xinhe North Star Delta
长沙北辰新河三角洲

Changsha | 长沙

Consisting of office, hotel and retail functions, the Changsha North Star Xinhe Delta Urban Complex, anchored by a 235-meter tower, was scheduled to be completed in an extraordinary 1,056 days, requiring standout efforts in overall organization, construction deployment and general contracting. (Please see page 66 for further information about the Changsha North Star Xinhe Delta project)

The project won the Quality Award in 2015 in Hunan province, National Quality Award in 2015, the Golden Award of China Steel Structure in 2015, National Method, and the Chinese Construction Enterprise Management Innovation of Science and Technology Award.

长沙北辰新河三角洲城市综合体包括办公区、酒店和零售等功能。中心是一座高达235 m的建筑。该项目施工时间只有短短的1 056天，对整体组织、施工部署及总承包商等方面提出了极高的要求。

项目最终获得2015年湖南省优质工程奖、2015年国家优质工程奖、2015年中国钢结构金奖，国家级工法1项，中国施工企业管理协会科学技术奖科技创新成果2等奖1项。

更多关于长沙北辰新河三角洲城市综合体的相关信息，请查阅第66页。

**Top:** Building under Construction
上图: 施工中的长沙北辰新河三角洲
**Bottom:** Overview of the building
下图: 外观全景

# China Tall Building Construction Award – Nominee
# 中国高层建筑建造奖提名作品

## Tianjin Kerry Center | 天津嘉里中心
Tianjin | 天津

The Tianjin Kerry Center is notable for several major construction achievements. Its foundation pit involves a combination of techniques that serve to counteract large deformations in soft soil due to underground water infiltration. The assessment and control mechanism for conducting the pit excavation was notable for its use of digital simulation technology prior to the dig, and for its integration with onsite monitoring during construction. A two-level welding and dismantling platform made of steel and composite board was deployed, accelerating construction speed, and a two-layer reinforcement bar connection was fashioned, offering added protection to the apartment areas.

天津嘉里中心因多项施工方面的成就而闻名。由于地下水渗透，地基坑内布满了软土，为了抵消由此产生的巨大形变，施工方对地基坑采用了一系列技术手段。对挖掘地基坑进行的评估与控制手段也值得一提。施工前采用了数字仿真技术，施工期间还随时对现场进行监测。为了加快施工速度，该项目采用了两级焊接技术加上由钢板和复合板造成的可拆解平台。此外还有一个双层加固钢筋连接构件，为公寓区提供额外的保护。

**Completion Date:** 2015
**Height:** 213 m (699 ft)
**Stories:** 61
**Owner:** Kerry Properties Limited; Shangri-La (Asia) Co., Ltd.; Allgreen Properties Limited
**Developer:** Tianjin Kerry Real Estate Development Co., Ltd.
**Main Contractor:** China Construction First Group Construction & Development Co., Ltd.

竣工时间：2015
高度：213 m（699 ft）
层数：61
业主：嘉里建设有限公司；香格里拉（亚洲）有限公司；长春产业有限公司
开发商：天津嘉里房地产开发有限公司
总承包商：中建一局集团建设发展有限公司

# China Outstanding Achievement Award Winner
中国高层建筑杰出贡献奖

# Awards Criteria
评选标准

This award recognizes an individual who has made extraordinary contributions to the advancement of tall buildings and the urban environment in China during his or her professional career. These contributions and leadership are recognized by the professional community and have significant effects, which extend beyond the professional community, to enhance cities and the lives of their inhabitants. The individual's contributions may be well known or little known by the public and may take any form, such as completed buildings, research, technology, methods, ideas, or industry leadership.

The candidate may be from any area of specialization, including, but not limited to: architecture, building systems, construction, academia, planning, development, or management. The award emphasizes the unique, multi-disciplinary nature of the CITAB and CTBUH, and is thus set apart from other professional organizations' awards for single disciplines.

1. The candidate must hold a citizenship of the Greater China region, including Hong Kong, Macau, Taiwan and Mainland China.

2. The personal attributes of the award recipient should be of high professional integrity and the individual's contributions generally consistent with the values and mission of the CITAB and CTBUH.

3. The candidate may or may not be a member of CITAB or CTBUH.

4. Awards are not normally given posthumously, but may be awarded to the deceased in unusual circumstances.

该奖项是表彰在职业生涯中为中国高层建筑以及城市环境发展作出杰出贡献的个人。他们的贡献以及领导力受到业内专业人士的认可并产生了重大影响，他们所作的贡献还延伸到专业领域之外，提高了城市以及当地居民的生活质量。他们个人的成就可能已经广为公众所知，也可能并不为人所知；成就也可以是多种形式的，例如：已竣工的建筑、研究、技术、方法、概念，或业内领导力。候选人可来自任何专业背景，包括但不限于：建筑设计、建筑系统、施工建造、学术研究、城市规划、项目开发或管理等。

该奖项强调 CITAB 和 CTBUH 所倡导的独特性和多学科性，因此奖项的设置有别于其他专业组织颁发的单一学科奖项。

1. 候选人必须为大中华地区人士，包括中国大陆和香港、澳门、台湾地区。
2. 候选人应具有较高的职业诚信与素养，且个人成就应与 CITAB 和 CTBUH 的价值观及使命相一致。
3. 候选人不要求一定为 CITAB 或 CTBUH 成员。
4. 已故人士一般不作为候选人。

# China Outstanding Achievement Award – Winner
## 中国高层建筑杰出贡献奖

## Dasui Wang | 汪大绥
East China Architectural Design & Research Institute (ECADI)
华东建筑设计研究总院

Left: CCTV Headquarters, Beijing, 2012 (234 m / 768 ft); the CTBUH Best Tall Building Worldwide 2013
左图：CCTV总部大楼，北京，2012 (234 m/768 ft)；2013年CTBUH世界最佳高层建筑
Above: Dasui Wang
上图：汪大绥

Dasui Wang was born in Leping, Jiangxi, in 1941. After his graduation from Tongji University in 1964, he devoted his career to structural design for 52 years and was awarded Engineering Design Master of China in 2000. Wang is currently the Chief Engineer for East China Architectural Design & Research Institute (ECADI), the Deputy Director of the Building Structure Committee of the Architectural Society of China and a member of the Expert Committee on Over-limit Tall Building Structures of the Ministry of Housing and Urban-Rural Development. He is also an adjunct professor and PhD supervisor at Tongji University.

In the first stage (1964-1978) of his lifelong design career, Wang worked at Lianyungang Institute of Architectural Design & Research, Jiangsu, mainly engaged in the structural design of industrial and civil buildings. As a young professional, he worked hard to study and improve technologies and made early achievements in promoting pre-stressed reinforced concrete structures, clay-based soil treatment, and

汪大绥，江西乐平人，生于1941年。1964年毕业于同济大学，从事结构设计52年，2000年被授予中国工程设计大师。现任华东建筑设计研究总院资深总工程师，同时兼任中国建筑学会高层建筑结构专业委员会副主任、住房和城乡建设部超限高层结构专家委员会委员，同济大学兼职教授、博士生导师。

汪大绥漫长设计生涯的前一段（1964—1978）在江苏省连云港市建筑设计院工作，主要从事各种类型的工业与民用

*"The modern history of Chinese tall building design is embodied in Dasui Wang, more than any other person alive today."*

"中国高层建筑设计的现代史在汪大绥身上得以具体体现，这超过了现今的任何人。"

Design & Technical Jury　评委会

lightweight roofing system design (including grid and long-span steel-wood composite structures). In 1977, he honored the first Science Conference of Lianyungang with his attendance.

Wang was then reassigned to East China Industrial Architectural Design & Research Institute (now ECADI). At that time, China was just beginning to "open up and reform", hastening an unprecedented boom in the construction industry. Meanwhile China's accelerating urbanization, as well as the sharp contrast between the scale of population and limits on available land, brought opportunities for the development of tall buildings. From the late 1970s to the early 1990s, Wang and his team at ECADI, relying on their own effort as well as lessons from Chinese peers, explored the theories and methodologies of high-rise structural design. On this basis, they designed Shanghai Huating Hotel, the Oriental Pearl Radio & TV Tower, Shanghai Guangming Building and Baosteel Command Center. Given differing project requirements, they explored new techniques, including the design method of shear walls with irregular openings, analysis and design of high-level deep transfer beams, and the mechanism and joint construction of outriggers. These research outcomes have been successfully applied to the above mentioned projects. Wang also participated in the development of the SPS series of structural analysis software based on Chinese computers.

After the 1990s, amidst continuing reform, the scaling up of construction, and the flow of foreign investment, international design firms entered the Chinese market. They brought advanced design concepts, new materials and technologies. Furthermore, the Chinese government carried out policies to facilitate cooperation between domestic and foreign design firms. The combination of these forces helped to lift China's construction industry to a higher level. During his design collaboration with foreign partners, Wang demonstrated great sincerity and with his peers, stressed mutual learning and active participation, aiming at continuous improvement.

### Jury Statement ｜ 评委会评语

The structural engineering work of Dasui Wang underlies the success of some of the most significant tall buildings in China, if not the world. Across a decades-long career, which began in China's isolation from the world and has smoothly transitioned into its rapid integration at the forefront of the global tall building industry, he has proven himself resourceful and original, and yet capable of absorbing the best practices of his peers with equanimity. His advice and counsel are treated with the utmost respect by his contemporaries and his inheritors; he responds with thoroughness and due reflection.

汪大绥的结构工程设计作品在中国绝对标志着最重要的一批高层建筑的成就。在几十年的职业生涯中，汪大绥见证了中国高层建筑从无到有，并成为全球高层建筑行业先锋的过程。他的设计丰富多变，富有创意，同时也能够坦然借鉴吸收同侪的经验，由此他出色地证明了自己的能力。他为人谦逊，勤于思考，善于指导，得到同龄人和年轻后辈们的一致敬重。

建筑的结构设计。作为一名年轻的技术人员，他努力钻研技术，在推广预应力钢筋混凝土结构、软土地基处理、轻型屋盖系统设计（包括网架与大跨度钢木组合结构）等方面取得了一定成绩。1977年光荣参加了连云港市第一届科学大会。

1979年汪大绥调入华东工业建筑设计院（今华东建筑设计研究总院）。这时适逢改革开放，给建筑业带来了前所未有的繁荣。中国城市化进程的加速和人多地少的国情也给高层建筑的发展带来了机遇。从上世纪70年代末到90年代初，汪大绥和他所服务的华东院主要依靠自己的力量、借鉴国内同行的经验探索高层建筑结构设计的理论和方法，在此基础上设计了上海华亭宾馆、东方明珠广播电视塔、上海光明大厦、宝钢生产指挥中心等高层建筑和构筑物。根据设计项目的需要，在不规则开洞剪力墙的设计方法、高位转换深梁的分析与设计、伸臂桁架的机理和节点构造等方面进行了一些探索和研究，并成功应用于上述工程的设计。同时参与了基于国产电子计算机的SPS系列结构分析软件的部分编制工作。90年代之后，随着改革开放的深入、建设规模的扩大和外资的进入，外国设计公司开始进入中国设计市场，他们带来了先进的设计理念、新的材料和技术，中国政府也制定了中外设计单位开展合作设计的相关政策，这对中国建筑业的发展和水平的提高起到了积极的作用。他在与外方合作设计的过程中采取了真诚合作，相互学习，积极作为，力求提高的态度，设计建成了一大批有重大影响的超高层建筑，如上海环球金融中心、中央电视台总部大楼、天津津塔、天津高银117大厦、东方之门等。在这些项目的实施过程中，依托国内力量，组织了对项目中关键技术难点的攻关，其中包括高含钢率SRC柱承载力和延性的研究、复杂蝶形节点受力性能研究、考虑屈曲后效应钢板剪力墙受力性能研究及改进，BRB斜撑在超高层建筑中的应用研究等。这些研究成果不仅保证了工程的顺利建成，也有力地推动了我国建筑结构技术水平的提高。

汪大绥多年来一直力推结构原创，他认为只有原创才能更有效地培育新人，才能使我国由高层建筑大国变成高层建筑技术上的强国。在他的推动和引领下，近年来华东院已经完成了多栋400~500 m高度的超高层建筑及复杂连体高层建筑的原创结构设计，如武汉中心、大连绿地中心、南京金鹰天地广场等。多种类型的消能减震技术开始在这些项目中应用，并且通过设计实践培养出了一个比较雄厚的、有创造力的、年轻化的技术梯队。他认为，在这些青年骨干身上承载着中国高层建筑技术发展的未来。

汪大绥一直追求结构设计的完美。他认为完美的结构除了应符合国家建设方针、与建筑设计成为和谐的有机体之外，结构本身也应完美。他认为完美的结构首先应该是力学上合理的，即具有明确的力学概念，简捷的传力路线，稳定的结构形体。在实际工程中这一点往往受到其他条件的制约，但也应该在可能条件下努力追求。其次，结构构

Top Left: The New CCTV Headquarters SRC Column Experiment
上左图：CCTV总部大楼主楼SRC柱试验
Top Middle: Tianjin World Financial Center Steel Plate Wall Experiment
上中图：天津津塔钢板剪力墙试验
Top Right: Goldin Finance 117 Mega BRB)
上右图：天津高银117大厦巨型BRB支撑施工

Together they accomplished a number of influential tall buildings, such as Shanghai World Financial Center, the new CCTV Headquarters, Tianjin World Financial Center, Goldin Finance 117, Gate to the East, and more. During the implementation of these projects, key technical difficulties were tackled with Chinese resources. The challenges included bearing capacity and ductility of high steel-content reinforced concrete, understanding the mechanical behavior of complex butterfly joints, researching and improving the mechanical behavior of steel-plate shear walls, considering the post-buckling effect, and research of the This really needs to be spelled out and defined. Send these pages back to ECADI for urgent review. in supertall buildings. These research results not only guaranteed the success of the projects, but facilitated the improvement of China's building structure technologies overall.

**Top Left:** Shanghai Huating Hotel
上左图：上海华亭宾馆
**Top Middle:** Oriental Pearl Radio & TV Tower
上中图：东方明珠广播电视塔
**Top Right:** Shanghai World Financial Center
上右图：上海环球金融中心
**Bottom Left:** Tianjin Word Financial Center
下左图：天津津塔
**Bottom Middle Left:** Goldin Finance 117
下中左图：天津高银117大厦
**Bottom Middle Right:** Gate to the East
下中右图：东方之门
**Bottom Right:** Wuhan Center
下右图：武汉中心
**Bottom Left (Opposite Page):** Nanjing Golden Eagle International Plaza
对侧左图：南京金鹰天地广场
**Bottom Right (Opposite Page):** Dalian Greenland Center
对侧右图：大连绿地中心

Wang has always insisted on original structural design. He believes that only through originality can we cultivate new talents and let China pride itself on the technological superiority, instead of the mere quantity, of tall buildings. In recent years, under Wang's lead, a number of original structural designs of some supertall buildings and connected towers, such as Wuhan Center, Dalian Greenland Center, Nanjing Golden Eagle International Plaza, etc., have been completed. Various types of energy dissipation technologies have been applied in these projects, and during the implementation process, a promising and creative team of young professionals has formed. Wang believes that these top talents carry on their shoulders the future of China's tall building technology, and that his true legacy will be in the works of the next generation. For these pioneers, he has the following advice.

Wang has always pursued perfection in structural design. In his opinion, a perfect structure should not only comply with state policies on construction and form harmonious synergy with architectural design, but should be perfect *per se*. He thinks that a perfect structure should first of all be plausible in terms of mechanics, i.e. have clearly identified mechanical concepts, short and direct force transmission paths, and a stable structure and form. In actual engineering situations, this ideal is often restricted by other conditions. But still, designers should push for it whenever possible.

Secondly, the layout of structural elements should achieve as graceful a composition as possible, i.e. the general principles of aesthetics should also be embodied in structural layout. Lastly, attention should also be paid to elegance of details. The relationship between components and arrangement of nodes should be carefully considered.

"Don't make any hasty decisions that you'll regret," Wang has often said. "We are no longer in an era of rough and massive production; people's pursuit of beauty should be demonstrated in the creations of structural engineers."

Wang believes that structure is the physical carrier on which a building and its functions rely. As projects that consume tremendous amounts of wealth, building structures must withstand many challenges from the natural environment during their long lives. Besides, they are essential to public security and social development. These features require that the structural designers should work diligently and take ultimate responsibility for their designs. Structural engineers should also keep learning throughout their lifetime, continuously hyphenate if splitting the word and applying new technologies to structural design, so as to improve the technological level of structural engineering and align structural design with the requirements of social development.

件的布置应尽量符合构图美，即美学的一般原则也应尽量在结构布置中体现。最后，还要注意细部美，构件相互关系、节点的交待都应琢磨，不要草率从事而留下遗憾。我们已经跨越了粗制滥造的年代，人民对美的追求也应该体现在结构工程师的创作中。

汪大绥认为结构是建筑赖以存在并实现其功能的物理载体。建筑结构消耗巨大的社会财富、经受各种自然因素严苛的考验，使用寿命很长，并且与社会发展和公众安全密切相关。这些特点要求结构设计人员时刻牢记自己责任的重大，必须兢兢业业工作，以对自己的设计终生负责。结构工程师还要终生学习，不断地把科学技术发展的新成果转化、应用到结构设计中去，以提高结构工程的技术水平，满足社会发展对结构设计的需要。

# 2016 年中国高层建筑奖评审委员会

2016 年中国高层建筑奖评审委员会是由世界高层建筑与都市人居学会（CTBUH）和中国高层建筑国际交流委员会（CITAB）从各自的支持网络中邀请并联合委任的高层建筑领域的专家。

中国高层建筑奖评审委员会负责评选"中国最佳高层建筑奖"的入围和获奖作品以及"中国高层建筑城市人居奖"、"中国高层建筑建造奖"和"中国高层建筑创新奖"的获奖作品与荣誉奖获奖作品。CITAB 理事会和 CTBUH 中国办公室董事会负责评选"中国高层建筑杰出贡献奖"获奖者和"中国高层建筑成就奖"的获奖作品与荣誉奖获奖作品。

**评委会会议报道**

2016 年 1 月 21 日，经过几个月的酝酿和筹备，首届中国高层建筑奖评选活动专家评审会在上海举行。

中国高层建筑国际交流委员会（CITAB）和世界高层建筑与都市人居学会（CTBUH）于 2015 年 10 月联合启动了首届中国高层建筑奖评选活动。自活动正式启动以来，得到了中国建筑学会和上海市建筑学会及高层建筑领域各界人士的鼎力支持，提交作品之多样以及支持者之众多体现了近年来高层建筑在中国蓬勃发展的趋势。

至 2015 年 12 月 14 日申报截止日期为止，主办方共计收到各类奖项申报作品近 90 个，申报项目为近两年来竣工的高层建筑，项目所在地遍及中国大陆和港澳台地区。中国境内公司约占 45%，境外公司约占 55%，其中包括各大海内外知名开发商和设计事务所。作为该奖项的首届活动，活动的受欢迎程度充分体现了其在高层建筑领域的广泛影响力。本次参评的作品不仅有来自北京、上海、广州、深圳等一线城市的高层建筑，还有来自天津、重庆、杭州、南京、武汉、长沙、郑州等城市的建筑。

在评审工作开始之前，中国高层建筑国际交流委员会（CITAB）主席、华东建筑设计研究总院院长张俊杰先生，世界高层建筑与都市人居学会（CTBUH）主席、KPF 建筑事务所合伙人 David Malott 先生，CTBUH 执行理事长 Antony Wood 博士，中国建筑学会副理事长周畅先生和上海市建筑学会理事长曹嘉明先生进行了简短的致辞，并介绍了本次评审会的评委。

# 2016 China Awards Jury

The 2016 China Awards Jury consists of experts in the tall building field are jointly appointed by CTBUH and CITAB from within their respective member networks.

The China Awards Jury is responsible for selecting the winners and honorable distinction recipients of the China Best Tall Building, Urban Habitat, Construction, and Innovation awards. The CITAB Board of Trustees and the CTBUH China Office Board are responsible for selecting the China Tall Building Outstanding Achievement Award and China Tall Building Legacy Award recipients.

## Jury Meeting Report

SHANGHAI – On January 21, 2016, the inaugural China Tall Building Awards jury meeting was held in Shanghai, the culmination of many months of effort.

The China International Exchange Committee for Tall Buildings (CITAB) and the Council on Tall Buildings and Urban Habitat (CTBUH) jointly launched the inaugural China Tall Building Awards in October 2015. Early on, the program received the kind support of the Architectural Society of China, the Architectural Society of Shanghai and individuals in the tall building industry — a diversity of supporters and submitters that reflect China's booming trend in tall building construction in recent years.

As of December 14th, 2015, the application submission deadline, the organizers have received nearly 90 submittals, which spanned projects completed in the past two years and spread out across the Chinese mainland, Hong Kong, Macau and Taiwan. Among the submittals, about 45% were submitted by Chinese companies, while the overseas companies constituted about 55% of the submittals. Many of the submittals were from were well-known developers and famous architects, fully embodying the event's wide range of influence in the high-rise industry, even in its first year. The projects considered were not only from Tier 1 cities, such as Beijing, Shanghai, Guangzhou and Shenzhen; cities including Tianjin, Chongqing, Hangzhou, Nanjing, Wuhan, Changsha and Zhengzhou all fielded entries.

Prior to the deliberations, Junjie Zhang, CITAB/ECADI Chairman, Antony Wood, CTBUH Executive Director, David Malott, CTBUH Chairman & Principal at KPF, and Chang Zhou, Vice Chairman of the Architectural Society of China, and Jiaming Cao, Chairman of the Architectural Society of Shanghai, gave a few words of encouragement. The jurors were then introduced:

设计评委：

主席：宋春华，中国住房和城乡建设部原副部长

周畅，中国建筑学会副理事长

曹嘉明，上海市建筑学会理事长

程泰宁，中国工程院院士

Stefan Krummeck，Farrells 事务所设计总监

David Malott，世界高层建筑与都市人居学会 (CTBUH) 主席、KPF建筑事务所合伙人

Antony Wood，世界高层建筑与都市人居学会 (CTBUH) 执行理事长

张俊杰，中国高层建筑国际交流委员会 (CITAB) 主席、华东建筑设计研究总院院长

吴长福，同济大学建筑与城市规划学院教授、同济大学建筑设计研究院(集团)有限公司副总裁

陈建邦，瑞安房地产发展有限公司发展及设计总监

技术评委：

江欢成，中国工程院院士

李国强，同济大学土木工程学院教授

龚剑，上海建工集团股份有限公司总工程师

顾建平，上海中心大厦建设发展有限公司总经理

Abdo Kardous，Hill International 资深副总裁

Michael Kwok，奥雅纳（上海）董事/总经理

谢锦泉，科进|柏诚中国区建筑机电董事总经理

朱毅，Thornton Tomasetti 资深董事、上海办公室总监

## Design Jury

**Chair: Chunhua Song,** former Vice Minister, Chinese Ministry of Housing and Urban Construction

**Chang Zhou,** Vice Chairman, Architectural Society of China

**Jiaming Cao,** Chairman, Architectural Society of Shanghai China

**Taining Cheng,** Academician, Chinese Academy of Engineering

**Stefan Krummeck,** Design Principal, Farrells

**David Malott,** CTBUH Chairman and Principal, Kohn Pedersen Fox Associates

**Antony Wood,** CTBUH Executive Director; Research Professor, Illinois Institute of Technology; and Visiting Professor, Tongji University

**Junjie Zhang,** Chairman, CITAB and ECADI

**Changfu Wu,** Professor, Architecture and Urban Planning of Tongji University and Vice President, Tongji Architecture Design (Group) Co.

**Albert (Jianbang) Chan,** Director of Development, Planning and Design, Shui On Development, Ltd.

## Technical Jury

**Huancheng Jiang,** Academician, Chinese Academy of Engineering

**Guoqiang Li,** Professor, Structural Engineering, College of Civil Engineering, Tongji University

**Jian Gong,** Chief Engineer, Shanghai Construction Group

**Jianping Gu,** General Manager, Shanghai Tower Construction & Development Co. Ltd.

**Abdo Kardous,** Senior Vice President and Managing Director, Asia-Pacific Operations, Hill International

**Michael Kwok,** Director, Arup (Shanghai)

**Vincent Tse,** Managing Director, Building MEP China, WSP | Parsons Brinckerhoff

**Yi Zhu,** Managing Principal, Shanghai Office Director, Thornton Tomasetti

# Index of Buildings | 建筑索引

5 Corporate Avenue, *Shanghai*;
企业天地中心5号楼，上海 — 60

**A**

Agile Center, *Guangzhou*;
雅居乐中心，广州 — 66

Asia Pacific Tower & Jinling Hotel, *Nanjing*;
金陵饭店亚太商务楼，北京 — 3

**B**

Bank of China Tower, *Hong Kong*;
中国银行大厦，香港 — 143

Bank of China Tower, *Shanghai*;
上海中银大厦，上海 — 143

Beijing Greenland Center, *Beijing*;
北京绿地中心，北京 — 170

Bund SOHO, *Shanghai*;
外滩SOHO，上海 — 9

**C**

CCTV Headquarters, *Beijing*;
CCTV总部大楼，北京 — 177

Center 66, *Wuxi*;
无锡恒隆广场，无锡 — 98

Central Plaza, *Hong Kong*;
香港中环广场，香港 — 142

Changsha North Star Xinhe Delta, *Changsha*;
长沙北辰新河三角洲，长沙 — 68, 172

Changzhou Modern Media Center, *Changzhou*;
常州现代传媒中心，常州 — 70

China Resources Building, *Hong Kong*;
华润大厦，香港 — 140

Chongqing Land Group Headquarters, *Chongqing*;
重庆地产集团总部，重庆 — 72

Chongqing Raffles City, *Chongqing*;
重庆来福士广场，重庆 — 158

CITIC Plaza, *Guangzhou*;
广州中信广场，广州 — 146

Colorful Yunnan • Flower City, *Kunming*;
七彩云南花之城，昆明 — 98

Corporate Avenue 6, 7 & 8, *Chongqing*;
企业天地6, 7, 8号楼，重庆 — 99

**D**

Dalian World Trade Center, *Dalian*;
大连世贸大厦，大连 — 146

Dachong Commercial Center, *Shenzhen*;
大涌商务中心，深圳 — 99

Ding Sheng BHW Taiwan Central Plaza, *Taichung*;
鼎盛BHW台湾中心广场，台中 — 100

**E**

East China Electrical Power Distribution Building, *Shanghai*;
华东电力调度大楼，上海 — 134

Evergrande Huazhi Plaza, *Kunming*;
恒大华置广场，成都 — 100

**F**

Fake Hills Linear Tower, *Beihai*;
假山大厦，北海 — 27

Forebase Financial Plaza, *Chongqing*;
申基金融广场，重庆 — 74

Fortune Financial Center, *Beijing*;
北京财富金融中心，北京 — 76

Forum 66, *Shenyang*;
沈阳市府恒隆广场，沈阳 — 167

Fuzhou Shenglong Financial Center, *Fuzhou*;
福州升龙汇金中心，福州 — 101

**G**

Global Harbor, *Shanghai*;
环球港，上海 — 101

Grand Hyatt Dalian, *Dalian*;
大连君悦酒店，大连 — 102

Guangming Building, *Shanghai*;
光明大厦，上海 — 147

Guangzhou Development Center Building, *Guangzhou*;
广州发展中心大厦，广州 — 144

**H**

Hangzhou Citizen Center, *Hangzhou*;
杭州市民中心 — 163

Heart of Lake, *Xiamen*;
万科湖心岛，厦门 — 117

Hong Kong and Shanghai Bank, *Hong Kong*;
汇丰银行，香港 — 131

Hongkou SOHO, *Nanjing*;
虹口SOHO，上海 — 15

Hopewell Center, *Hong Kong*;
合和中心，香港 — 145

Hua Nan Bank Headquarters, *Taipei*;
华南银行总部大楼，台北 — 31

Huizhou Central Place, *Huizhou*;
惠州华贸中心，惠州 — 124

HVW Headquarters, *Taoyuan*;
台湾HVW总部，桃园 — 78

**I**

International Foreign Trade Center, *Shenzhen*;
国贸中心，深圳 — 132

**J**

J57 SkyTown, *Changsha*;
J57天空之城，长沙 — 80

Ji'nan Greenland Center, *Jinan*;
济南绿地中心，济南 — 82

Jing An Kerry Center, *Shanghai*;
静安嘉里中心，上海 — 111

Jing Guang Center, *Beijing*;
京广中心，北京 — 142

Jing Mian Xin Cheng Tower, *Beijing*;
京棉新城大厦，北京 — 84

Jin Jiang Tower Hotel, *Shanghai*;
上海新锦江大酒店，上海 — 141

Jin Mao Tower, *Shanghai*;
金茂大厦，上海 — 137

JW Marriott Shenzhen Bao'an, *Shenzhen*;
深圳前海华侨城JW万豪酒店，深圳 — 102

**K**

Kingtown International Center, *Nanjing*;
南京金奥国际中心，南京 — 86

Kunlun Hotel, *Beijing*;
昆仑饭店，北京 — 140

## L

Lippo Center, *Hong Kong*;
力宝中心, 香港 — 141

Lujiazui Century Financial Plaza, *Shanghai*;
陆家嘴世纪金融广场, 上海 — 43

## M

Macau Tower, *Macau*;
澳门旅游塔, 澳门 — 147

Mount Parker Residences, *Hong Kong*;
西湾台1号, 香港 — 103

## N

Nanchang Greenland Central Plaza, *Nanchang*;
南昌绿地中心, 南昌 — 35, 154

Nanchang Greenland Zifeng Tower, *Nanchang*;
南昌绿地紫峰大厦, 南昌 — 39

Nanjing Zifeng Tower, *Nanjing*;
南京紫峰大厦, 南京 — 162

Ningbo Global Shipping Plaza, *Ningbo*;
宁波环球航运广场, 宁波 — 103

## O

OLIV, Hong Kong;
香港OLIV, 香港 — 47

Oriental Blue Ocean International Plaza, *Shanghai*;
东方蓝海国际广场, 上海

Oriental Financial Center, *Shanghai*;
东方汇经中心, 上海 — 88

## P

People's Daily New Headquarters, *Beijing*;
人民日报新总部, 北京 — 51

Pudong International Information Port, *Shanghai*;
浦东国际信息港, 上海 — 147

## R

R&F Yingkai Square, *Guangzhou*;
富力盈凯广场, 广州 — 90

## S

Shanghai Arch, *Shanghai*;
上海金虹桥国际中心, 上海 — 92

Shanghai Center, *Shanghai*;
上海商城, 上海 — 135

Shanghai Tower, *Shanghai*;
上海中心大厦, 上海 — 151

Shanghai World Financial Center, *Shanghai*;
上海国际金融中心, 上海 — 178

Shaoxing Shimao Crowne Plaza, *Shaoxing*;
绍兴世茂皇冠假日酒店, 绍兴 — 104

Shenye Tairan Building, *Shenzhen*;
深业泰然大厦, 深圳 — 121

Shenzhen Development Bank, *Shenzhen*;
深圳发展银行大厦, 深圳 — 145

Shenzhen Xinhe World Office, *Shenzhen*;
深圳星河World写字楼, 深圳 — 104

Shenzhen Zhongzhou Holdings Financial Center, *Shenzhen*;
深圳市中洲控股金融中心, 深圳 — 94

Shun Hing Square, *Shenzhen*;
信兴广场, 深圳 — 136

Studio City, *Macau*;
新濠影汇, 澳门 — 105

Sunrise Kempinski Hotel, *Beijing*;
日出东方凯宾斯基酒店, 北京 — 96

## T

TAIPEI 101, *Taipei*;
台北101大楼, 台北 — 139

Taiping Finance Tower, *Shenzhen*;
深圳太平金融大厦, 深圳 — 55

T & C Tower, *Kaohsiung*;
高雄85大楼, 高雄 — 143

The Center, *Hong Kong*;
香港中环中心, 香港 — 146

The Landmark Gloucester Tower, *Hong Kong*;
告罗士打大厦, 香港 — 145

The Wave of Science and Technology Park S01, *Jinan*;
浪潮科技园S01科研楼, 济南 — 105

Tianjin International Trade Tower 1, 2 & 3, *Tianjin*;
天津国际贸易中心1, 2, 3号楼, 天津 — 106

Tianjin Kerry Center, *Tianjin*;
天津嘉里中心, 天津 — 172

Tomorrow Square, *Shanghai*;
明天广场, 上海 — 144

Two International Finance Center, *Hong Kong*;
国际金融中心二期, 香港 — 138

## W

Wangjing SOHO, *Beijing*;
望京SOHO, 北京 — 21, 173

White Swan Hotel, *Guangzhou*;
白天鹅宾馆, 广州 — 130

WPP Campus, *Shanghai*;
达邦协作广场, 上海 — 106

Wuhan Center, *Wuhan*;
武汉中心, 武汉 — 156, 160

Wuhan Tiandi Site A, *Wuhan*;
武汉天地A座, 武汉 — 126

Wuxi Suning Plaza 1, *Wuxi*;
无锡苏宁广场, 无锡 — 107

## X

Xiamen World Overseas Chinese International Conference Center, *Xiamen*;
厦门世侨中心, 厦门 — 107

## Z

Zhengzhou Greenland Plaza, *Zhengzhou*;
郑州绿地中心·千玺广场, 郑州 — 59

# Index of Companies ｜ 企业索引

**A**

ACLA; 傲林国际设计有限公司 66, 74, 90
Adrian L. Norman Limited; Adrian L. Norman有限公司 103
Adrian Smith + Gordon Gill Architecture; 芝加哥Adrian Smith + Gordon Gill 建筑事务所 94, 95
AECOM; 艾奕康建筑设计有限公司 100, 104, 105, 136, 142, 146
Aedas; 凯达环球 98, 100
Agile Group; 雅居乐集团 65
AIM Architecture; 恺慕建筑 9
Allgreen Properties Limited; 长春产业有限公司 173
ALT Limited; ALT 有限公司 92, 98
American Design Associates; 美国设计有限公司 136
Aon Fire Protection Engineering; Aon消防工程 35
Architecture and Engineers Co., Ltd. of Southeast University; 东南大学建筑设计研究院 51
Arnold Associates; Arnold设计公司 92
Arquitectonica; Arquitectonica建筑设计事务所 103
Artmost Building Decoration Design Engineering Co., Ltd.; Artmost建筑装饰设计工程有限公司 163
Arup; 奥雅纳工程咨询有限公司 21, 62, 74, 90, 103, 105, 106, 111, 131, 138, 142, 144, 145, 158
Associated Consulting Engineers; Associated工程顾问公司 134, 136, 146
Aurecon; 澳昱冠工程咨询有限公司 100

**B**

Bank of China; 中国银行 134
Beca Group; 贝科工程设计集团 147
Beihai Xinpinguangyang Real Estate Development Co. Ltd.; 北海馨平广洋房地产开发有限公司 27
Beijing Construction Engineering Group; 北京建工集团 66
Beijing Enterprises International Conference Metropolis Real Estate Co., Ltd.; 北京北控国际会都房地产开发有限公司 96
Beijing Institute of Architectural Design; 北京市建筑设计研究院 51, 66, 70, 83, 94, 105
Beijing R&F Properties Development Co.; 北京富力地产股份有限公司 90
Beijing Rundexin International Resources Investment Co., Ltd.; 北京润德信国际资源投资有限公司 170
Beijing Shenglong Electric Equipment Co., Ltd.; 北京盛隆电气有限公司 170

Beijing Urban Construction Group Co., Ltd.; 北京城建集团有限责任公司 96
Benoy; 英国贝诺建筑设计公司 100, 111
Benway Limited; Benway有限公司 47
Benwood Studio Shanghai; 本杰明·伍德建筑事务所 126
B.M. Holding (Group) Co., Ltd.; 宝矿控股(集团)有限公司 106
BPI (Brandston Partnership, Inc.); 美国碧谱照明设计有限公司 84, 90, 98, 99, 100
Broad Sustainable Building Co., Ltd.; 远大可建科技股份有限公司 78
Buro Happold; 英国标赫工程顾问公司 103

**C**

Campbell Shillinglaw Lau Ltd.; 金宝声学顾问公司 3
CapitaLand Limited; 凯德置地 106
CCDI Group; 悉地国际 21, 43, 101
CCM Architects; CCM建筑事务所 147
Celia Chu Design; 102
Central Southern Geotechnical Design Institute Co., Ltd.; 中南勘察设计院股份有限公司 78
CERI, Ltd.; 中冶京诚工程技术有限公司 74
Cesar Pelli & Associates; 西萨·佩里建筑设计事务所 138
Changzhou Broadcasting Station; 常州广播电台 68
Changzhou Radio and TV Realty Company, Ltd.; 常州广播电视实业有限公司 68
Chapman Taylor; 查普门·泰勒建筑设计事务所 101
Cheng Yuan MEP Consultants; Cheng Yuan机电咨询公司 76
Cheung Kong Holdings; 长江实业 88, 146
Chhada Siembieda Leung Ltd.; Chhada Siembieda Leung 有限责任公司 3
Chien Tai Cement Corporation; 建台水泥公司 143
China Academy of Building Research; 中国建筑科学研究院 21, 121
China Architecture Design & Research Group; 中国建筑设计研究院 102
China Construction Eighth Engineering Division Corp. Ltd.; 中国建筑第八工程局有限公司 88, 99
China Construction First Building (Group) Corporation Limited; 中国建筑一局(集团)有限公司 98, 102, 104
China Construction First Group Construction & Development Co., Ltd.; 中建一局集团建设发展有限公司 107, 173

China Construction Fourth Engineering Division Corp. Ltd.; 中国建筑第四工程局有限公司 99, 147
China Construction Steel Structure Corporation Ltd.; 中建钢构有限公司 167
China Construction Third Engineering Bureau Co., Ltd.; 中国建筑第三工程局有限公司 55, 68, 99, 106, 124, 126, 143, 156, 167
China International Trust and Investment; 中国国际信托投资公司 146
China Jin Mao Group Co., Ltd.; 中国金茂(集团)股份有限公司 137
China Railway Construction Engineering Group; 中国铁路工程总公司 94
China Resources (Holdings) Company Limited; 华润(集团)有限公司 140
China Resources Land Limited; 华润置地有限公司 99, 102
China Shipbuilding NDRI Engineering Co., Ltd.; 中船第九设计研究院工程有限公司 101
China Southwest Design & Research Institute; 中国建筑西南设计研究院有限公司 100
China State Construction Engineering Corporation; 中国建筑工程总公司 3, 21, 86, 100, 107, 117, 136
China Taiping Insurance Group Ltd.; 中国太平保险集团有限责任公司 55
Chongqing Architecture and Design Institute; 重庆市设计院 72
Chongqing Land Group; 重庆地产集团 70
Chroma33 Architectural Lighting Design; 大公照明设计顾问有限公司 31
Chuang Qing Facade Consultant; 创青幕墙顾问公司 103
Chun Cheng Construction; 春成建筑公司 76
Chyi Yuh Construction; Chyi Yuh建筑公司 100
Cleveland Bridge Ltd.; 克利夫兰桥梁建筑公司 131
Continental Engineering Consultants, Inc.; Continental工程顾问有限公司 139, 143
CR Construction; 华润建筑有限公司 140
CS Caulkins Co. Inc; CS Caulkins有限公司 84
C.Y. Lee & Partners Architects/Planners; 李祖原联合建筑师事务所 139, 143, 147

**D**

Da Join Architects & Associates; Da Join建筑设计公司 100
Dennis Lau & Ng Chun Man Architects & Engineers

(HK) Ltd. (DLN); 刘荣广伍振民建筑师事务所有限公司 101, 103, 142, 146

Design Land Collaborative; 地茂景观设计咨询（上海）有限公司 62, 88, 98, 126

## E

Earth Asia; 泛亚国际 105

ECADI (Shanghai East China Architecture Design and Research Institute Co., Ltd.); 华东建筑设计研究总院 9, 39, 43, 59, 62, 80, 86, 88, 92, 104, 124, 133, 135, 156, 160, 162, 177

Ecoland; 易兰 21

Edgett Willams Consulting Group Inc.; Edgett Willams咨询集团 35, 39, 84

EHS ArchiLab + Hsuyuan Kuo Architects & Associates; 大尺设计+郭旭原建筑师事务所 76, 100

Elmich (Guangzhou) Landscape Engineering Co., Ltd.; Elmich（广州）景观设计有限公司 72

E Man-Sanfield JV Construction; 138

Environmental Market Solutions, Inc.; EMSI 21, 74, 99, 106

Evergrande Real Estate; 恒大地产 100

Evergreen Consulting Engineering; 永竣工程顾问股份有限公司 139, 143

## F

Food Service Consultants, Ltd.; 富思餐饮服务顾问有限公司 102

Forebase Group; 甲基集团 72

Fortune Consultants, Ltd.; 财富顾问公司 92

Foster Associates; 福斯特建筑事务所 131

## G

Gammon Construction Limited; 金门建筑有限公司 146

GDIL Lighting Design; 香港大观国际设计咨询有限公司 106

General Office of People's Daily New Headquarters Construction & Development Program; 人民日报新总部建设和发展项目办公室 51

Gensler; 晋思 106, 151

Goddard Group; Goddard设计集团 105

Goettsch Partners; 美国GP建筑事务所 90, 102

Great Earth Architects & Engineers International; 大地建筑事务所 78

Greenland Group; 绿地集团 39, 43, 59, 80, 170

Guangzhou Design Institute; 广州市设计院 64

Guangzhou Di-Er Construction & Engineering Co., Ltd.; 广州第二建筑工程有限公司 130

Guangzhou Residential Architectural Design Institute; 广州市住宅建筑设计院 90

Guangzhou R&F Properties Co., Ltd.; 广州富力地产股份有限公司 90

## H

HaiPo Architects; HPA海波建筑设计 43

Hang Lung Properties; 恒隆地产 98, 167

Harbour Century Limited; Harbour Century有限公司 47

HASSELL; HASSELL设计公司 106

Henan Guoji Gongcheng Jianshe Gufen Co. Ltd.; 河南国基工程建设股份有限公司 101

Henderson Land Development; 恒基兆业地产有限公司 138

High Wealth Construction; 兴富发建设 100

Hill Landscape Design; Hill景观设计公司 121

Hip Hing Construction; 协兴建筑有限公司 141

Hirsch Bedner Associates; HBA设计工程顾问有限公司 102, 111

HOK; HOK建筑师事务所 143

Hong Kong Construction (Holdings) Limited; 香港建设(控股)有限公司 134, 136, 142, 146

Hong Kong Land limited; 香港置地 145

Hsin Chong Group; 新昌集团 103

Hua Nan Commercial Bank Ltd.; 华南商业银行股份有限公司 31

Huasen Architectural & Engineering Designing Consultants Ltd.; 深圳华森建筑与工程设计顾问有限公司 102

Huizhou City Run and the Real Estate Development Co., Ltd.; 惠州市华贸兴业房地产有限责任公司 124

## I

Ikonik; 依科 21

Illumination Physics; 105

I.M. Pei & Partners; 贝聿铭建筑设计事务所 134

Inhabit Group; 英海特工程咨询集团 21, 99, 105, 106

Inspur Group; 浪潮集团 105

Institute of Shanghai Architectural Design & Research Co., Ltd.; 上海建筑设计研究院有限公司 68, 101, 111, 137, 143, 144, 147

ISSEY; 106

## J

Jaros, Baum & Bolles; JB&B工程师事务所 134

JDC Global Inc.; JDC Global有限公司 86

Jerde Partnership; 捷德国际建筑师事务所 66

Jhih Jhan MEP Consultants; Jhih Jhan机电工程顾问公司 100

Jiang Architects and Engineers; 江欢成建筑有限公司 27, 107

JiangSu Goldenland Real Estate Development Co., Ltd.; 江苏金大地(集团)房地产开发有限责任公司 84

Jiangsu Provincial Architectural D&R Institute Ltd.; 江苏省建筑设计研究院有限公司 3, 107

Jinan Inspur Mingda Information Technology Co., Ltd.; 济南浪潮铭达信息科技有限公司 105

Jinling Hotel Corporation Ltd.; 金陵饭店有限责任公司 3

John Portman & Associates; 约翰·波特曼建筑设计事务所 92, 102, 135, 144

J. Roger Preston Group; 澧信工程顾问公司 98, 103, 131, 138

Junefair Group; 增辉集团 140

## K

Kangtian Real Estate Co., Ltd.; 康田置业有限公司 70

Kaplan Gehring McCarrol Architectural Lighting, Inc; Kaplan Gehring McCarrol建筑照明设计公司 39, 43

Karbony Investment; Karbony投资公司 136

Ke Jian Structural Consultants; Ke Jian结构顾问公司 100

Kengo Kuma and Associates; 隈研吾建筑都市设计事务所 15

Kerry Properties Ltd., 嘉里建设有限公司 111, 173

King-Le Chang & Associates; 杰联国际工程顾问有限公司 31

Kohn Pedersen Fox Associates; KPF建筑事务所 63, 89, 111

Kris Yao | Artech; 姚仁喜 | 大元建筑工场 31

Kuang Chun Cheng Construction; 广春成建筑公司 76

Kumagai Gammon Joint Venture; 熊谷金门合资公司 146

189

Kumagai Gumi; 日本株式会社熊谷組 134, 136, 139

Kung & Lee Architects Ltd.; Kung & Lee建筑设计有限公司 134

K.Y. Cheung Design Associates; 张国言设计事务所 136

## L

Land Development Corporation; 土地发展公司 146

L&A Urban Planning and Landscape Design (Canada) Ltd.; 加拿大奥雅景观规划设计事务所 106

Lehr Associates; Lehr顾问公司 139

Leigh & Orange; 利安顾问有限公司 101, 105

Lerch Bates; Lerch Bates顾问公司 39, 43

Leslie E. Robertson Associates; 理雅结构工程咨询有限公司 134, 136, 138

Lightdesign; 21

Li Jin Engineering Co., Ltd.; 利晋工程股份有限公司 31

LTW Designworks; 102, 105

Lutron; 路创电子公司 72

Lu Wen Cheng Architects & Associates; Lu Wen Cheng建筑师事务所 76

## M

MAD Architects; MAD建筑事务所 27

Magnificence Interiors Inc.; 禾安室内装修设计工程股份有限公司 31

Manloze Ltd.; Manloze有限公司 142

MCC Huaye Resources Development Co., Ltd.; 中冶集团华冶资源开发有限责任公司 72

Meinhardt; 迈进工程设计咨询有限公司 3, 62, 88, 98, 102, 103, 105, 106, 142

Melco Crown Entertainment Limited; 新濠博亚娱乐有限公司 105

METROSTUDIO; 意大利迈丘设计 72

Mindscape Ltd.; Mindscape有限公司 103

Ming Shen Engineering Inc.; 茗生工程股份有限公司 31

MTR Corporation Limited; 香港铁路有限公司 138

MVA Transportation, Planning & Management Consultants; MVA弘达交通规划与管理顾问公司 103, 105

## N

Nadel; 美国纳德尔建筑设计公司 146

Nan Fung Group; 南丰集团 106

Nanjing Institute of Landscape Architecture Design & Planning Ltd.; 南京园林规划设计院有限公司 3

National Engineering Research Center for Fire Protection; 国家消防工程技术研究中心 78

Newcomb & Boyd; Newcomb & Boyd工程顾问公司 92, 144

New Jinling Hotel Limited Company; 新金陵饭店有限责任公司 3

New World Development Company Limited; 新世界发展有限公司 142

Nihon Sekkei; 株式会社日本设计 142

Nikken Sekkei Ltd.; 株式会社日建设计 35, 55, 103, 143, 147

Ningbo Global Properties Limited; 宁波环球置业有限公司 103

No.3 Construction Co., Ltd. of Chongqing Construction Engineering Group; 重庆建工集团第三建筑有限公司 70

North Star Real Estate Ltd.; 北辰房地产有限公司 66

## O

Olin Studio; 117

## P

Parsons Brinckerhoff Consultants Private Limited; 柏诚顾问有限公司 146

Paul Rudolph; 保罗·鲁道夫建筑事务所 141

Paul Y. Engineering Group; 保华建业 105

Paul Y - ITC Construction; 保华德祥建筑集团 146

Peddle Thorp & Walker; 柏涛建筑设计集团 145

Perkins Eastman; 美国Perkins Eastman建筑设计事务所 124

Perkins + Will; 帕金斯威尔建筑设计事务所 86

PLACEMEDIA; 31

P & T Group (Palmer & Turner Group); 巴马丹拿建筑设计咨询有限公司 3, 74, 99, 106, 126, 145

## R

RBS Architectural Engineering Design Associates; 广州容柏生建筑结构设计事务所 78, 102

Rider Levett Bucknall; 利比有限公司 98, 99, 102, 106

Robert A.M. Stern Architects; 罗伯特斯特恩建筑师事务所 117

Rocco Design Architects Limited; 许李严建筑师事务所有限公司 138

Rolf Jensen & Associates; 罗尔夫杰森消防技术咨询有限公司 45, 88, 101

Ronald Lu & Partners; 吕元祥建筑师事务所 140

RSEA Engineering; 荣工工程股份有限公司 139

RTKL; RTKL建筑事务所 62, 66, 107

Ruihua Construction; 深圳市瑞华建设股份有限公司 101

RWDI; 39, 43

## S

Samsung C&T Corporation; 三星C&T公司 139

Schindler; 迅达集团 101

Schmidlin; Schmidlin公司 74, 82

Shanghai Anlian Investment & Development Co.; 上海安联投资发展有限公司 144

Shanghai Citelum Lighting Design Co., Ltd.; 上海城市之光灯光设计有限公司 3

Shanghai Construction Group; 上海建工集团 80, 106, 111

Shanghai Construction No.1 (Group) Co., Ltd.; 上海建工一建集团有限公司 9, 15, 62, 120

Shanghai Construction No.2 (Group) Co., Ltd.; 上海建工二建集团有限公司 101, 144

Shanghai Huadu Architect Design Company; 上海华都建筑规划设计有限公司 96

Shanghai Information World Co., Ltd.; 上海信息世界有限公司 147

Shanghai Installation Engineering Co., Ltd.; 上海市安装工程集团有限责任公司 43

Shanghai Jin Hong Qiao International Property Co., Ltd.; 上海金虹桥国际置业有限公司 92

Shanghai Kighton Façade Consultants Co., Ltd.; 上海凯腾幕墙设计咨询有限公司 66

Shanghai Lujiazui Financial and Trade Zone Development Co., Ltd.; 上海陆家嘴金融贸易区开发有限公司 43

Shanghai Oriental Blue Ocean Real Estate Co., Ltd.; 上海东方蓝海置业有限公司 86

Shanghai Research Institute of Building Sciences; 上海建筑科学研究院 62

Shanghai Songer Lighting Design Co., Ltd.; 上海松尔照明设计有限公司 86

Shanghai Tomorrow Square Co., Ltd.; 上海明天广场有限公司 144

Shanghai Tower Construction & Development Co., Ltd.; 上海中心大厦建设发展有限公司 151

Shanghai Xian Dai Architecture Design (Group) Co., Ltd.; 上海现代建筑设计集团有限公司 106

Shanghai Xintiandi Management Limited; 上海新天地商业管理有限公司 62

Shanghai Xusheng Property Co., Ltd.; 上海旭升置业有限公司 15

Shangri-La (Asia) Co., Ltd.; 香格里拉(亚洲)有限公司 173

Shenglong Group; 升龙集团 101
Sheng Lue Architectural Technology Co., Ltd.; 盛略建筑科技有限公司 86
Shen Milsom Wilke, Inc.; 声美华顾问公司 31, 39, 43, 105
Shenzhen Catic Curtain Wall Engineering Co., Ltd.; 深圳中航幕墙工程有限公司 72
Shenzhen Dachong Industrial Co., Ltd.; 深圳市大冲实业股份有限公司 99
Shenzhen General Institute of Architectural Design and Research Co., Ltd.; 深圳建筑设计研究总院有限责任公司 55
Shenzhen Huayang International Engineering Design Co., Ltd.; 深圳市华阳国际工程设计股份有限公司 99
Shenzhen Investment Holdings Co.; 深圳投资控股有限公司 94
Shenzhen OCT Hotel Real Estate Co., Ltd.; 深圳市华侨城酒店置业有限公司 102
Shenzhen Property Development; 深房集团 132
Shenzhen Terra (Holdings) Co., Ltd.; 深圳泰然集团有限责任公司 121
Shenzhen Yabao Real Estate Development Co., Ltd.; 深圳市雅宝房地产开发有限公司 104
Shimao Group; 世茂集团 104
Shui On Land Limited; 瑞安集团 99, 126
Sino Land Company Limited; 信和置业有限公司 142
Sino-Ocean Land; 远洋地产 82
Siu Yin-Wai & Associates; 邵贤伟建筑工程师事务所 140
Skidmore, Owings & Merrill LLP; SOM建筑事务所 39, 43, 59, 65, 84, 126, 137, 155
Sky City Investment Co., Ltd.; 天空城市投资股份有限公司 78
Sociedade de Turismo e Diversoes de Macau; 澳门旅游娱乐股份有限公司 147
SOHO China Co., Ltd.; SOHO中国有限公司 9, 15, 21
Spark Architects; 思邦建筑 84, 106
Steve Leung Designers; 梁志天设计师有限公司 90
Studio Raymond Chau Architecture Limited; 周文渭建筑师事务所有限公司 47
Sun Hung Kai Properties Limited; 新鸿基地产集团有限公司 138, 142
Suning Real Estate Group; 苏宁地产集团 107
Super Potato; 90, 111
SWA Group; SWA 集团 39, 43, 84, 111
Swire Properties; 太古地产 103

T

Taipei Financial Center Corporation; 台北金融中心公司 139

Taiwan Kumagai; 华熊营造股份有限公司 139
Tai Yun Fa Structural Consultants; Tai Yun Fa结构咨询公司 76
Talent Mechanical and Electrical Engineers Ltd.; 汇智机电顾问有限公司 140
Tanghua Architect & Associates Co., Ltd.; 汤桦建筑设计事务所 72
Ta-You-Wei Construction; 大有为营造股份有限公司 139
The Architectural Design & Research Institute of Guangdong Province; 广东省建筑设计研究院 124
The Hong Kong and China Gas Company Limited; 香港中华煤气有限公司 138
Theo Kondos; 101
THEO TEXTURE; 47
Thornton Tomasetti; 宋腾添玛沙帝工程顾问公司 62, 139, 151
Tianjin Architecture Design Institute; 天津市建筑设计院 106
Tianjin Kerry Real Estate Development Co., Ltd.; 天津嘉里房地产开发有限公司 173
Tianyuan Construction Group Co., Ltd.; 天元建设集团 105
Tino Kwan Lighting Consultants Ltd.; 关永权照明设计有限公司 102, 105
Tongji Architectural Design (Group) Co., Ltd.; 同济大学建筑设计研究院(集团)有限公司 15, 98, 151, 163
Tridant Engineering Company Limited; Tridant工程有限公司 146
Trustful Engineering & Construction Co., Ltd.; 卓誉建筑工程有限公司 47
Tuntex Group; 东帝士集团 143
Turner International LLC; Turner国际建设公司 143
TY Lin International; 林同棪国际工程咨询有限公司 143

V

Valentine, Laurie, and Davis; 134
Vanke Group; 万科集团 117
von Gerkan, Marg and Partners Architects; GMP建筑师事务所 9, 144

W

WAA; 126
Watermark Associates; Watermark设计事务所 3
Weidlinger Associates; 威德林格工程师事务所 144
White Swan Hotel Group Co., Ltd.; 白天鹅酒店集团有限公司 130
William Tao & Associates, Inc.; William Tao工程顾问公司 143
WMKY Limited; 云麦郭杨建筑师工程师事务所 140, 145
Wong & Cheng Consultants Engineers Limited; 黄郑顾问工程有限公司 47
Wong & Ouyang; 王欧阳香港有限公司 111, 141
Wong Tung & Partners; 王董建筑师事务所有限公司 142
WSP Hong Kong Ltd.; 科进香港有限公司 145
WSP | Parsons Brinckerhoff; 科进 | 柏诚集团 74, 84, 100, 111
WT Partnership; 务腾咨询有限公司 74, 105
Wuhan Construction Engineering Group Co., Ltd.; 武汉建工(集团)有限公司 84

X

Xiamen BIAD Architectural Design Ltd.; 厦门佰地建筑设计有限公司 117
Xiamen Wocicc Investment Management Co., Ltd.; 厦门世侨投资管理有限公司 107
Xiang Jiang Xing Li Estates Development Ltd.; 香江兴利房地产开发有限公司 74
Xinxing Construction & Development Corporation; 新兴建设开发总公司 51

Y

Yau Lee Group; 有利集团 105
Yonsei University; 延世大学 21
Yuanda; 远大集团 82
Yuexing Group; 月星集团 101
Yunnan Mythic Flora Corporate; 云南怡美天香有限公司 98

Z

Zaha Hadid Architects; 扎哈·哈迪德建筑事务所 21
Zhongying Holding Group Co., Ltd.; 中盈控股集团有限公司 170
Zhejiang Zhong Tian Construction Group Co., Ltd.; 浙江中天建设集团有限公司 59
Zhongxin Architectural Design And Research Institute Pty. Ltd.; 中信建筑设计研究院有限公司 86
Zhubo Design; 筑博建筑设计集团有限公司 121

# Image Credits | 图片版权

**Front Cover:**

**Pg 2–7:** All © P & T Group

**Pg 8–13:** All photos © Christian Gahl; drawing © von Gerkan, Marg and Partners Architects

**Pg 14–19:** All © Tongji Architectural Design (Group) Co., Ltd

**Pg 20-21:** All © CCDI (Beijing) International Architectural Design Consultants Ltd.

**Pg 22:** Left © CCDI (Beijing) International Architectural Design Consultants Ltd; right © Zaha Hadid Architects

**Pg 23:** © CCDI (Beijing) International Architectural Design Consultants Ltd

**Pg 24–25:** All © Zaha Hadid Architects

**Pg 26–29:** All © MAD Architects

**Pg 30–33:** All © Kris Yao | Artech

**Pg 34–37:** All © Nikken Sekkei Ltd.

**Pg 38–40:** All © LV Hengzhong

**Pg 41:** Left © LV Hengzhong; right © Skidmore, Owings & Merrill LLP

**Pg 42–44:** © LV Hengzhong

**Pg 45:** Left © Skidmore, Owings & Merrill LLP; right © LV Hengzhong

**Pg 46–49:** All © Theo Texture

**Pg 50–53:** All © Architects and Engineers Associates | Southeast University

**Pg 54–57:** All © Greenwind Photo

**Pg 58:** © 高劲松

**Pg 60:** Left © Si-ye Zhang; right © Nacasa & Partners Inc.

**Pg 61:** © Skidmore, Owings & Merrill LLP

**Pg 62–63:** ECADI

**Pg 64:** © Skidmore, Owings & Merrill LLP

**Pg 65:** All © Guangzhou Design Institute

**Pg 66:** © RTKL | YiHuai Hu

**Pg 67:** Top left and right © RTKL; bottom © Beijing Institute of Architectural Design Co., Ltd.

**Pg 68–69:** All © Shanghai Institute of Architectural Design & Research

**Pg 70–71:** All © Beijing Institute of Architectural Design Co., Ltd

**Pg 72–73:** All © Tanghua Architects Shenzhen Co., Ltd.

**Pg 74–75:** All © P & T Group

**Pg 76–77:** All © Hsuyuan Kuo Architects & Associates

**Pg 78–79:** All © BROAD Group

**Pg 80–81:** All © ECADI

**Pg 82–83:** All © Spark

**Pg 84–85:** All © Nanjing Golden Land Group

**Pg 86:** © Mr. Zhong Hai Shen

**Pg 87:** Left © Mr. Zhong Hai Shen; top right © Perkins + Will; bottom right © James & Connor Steinkamp, Steinkamp Photography

**Pg 88–89:** All © ECADI

**Pg 90–91:** All © Goettsch Partners, Inc

**Pg 92–93:** Photos © Luo Wen/VMA VISUAL; drawings © John Portman & Associates

**Pg 94–95:** All © Beijing Institute of Architectural Design

**Pg 96–97:** All © Shanghai Huadu Architecture & Urban Planning Co. Ltd.

**Pg 98:** Left © Aedas; right © Tongji Architectural Design (Group) Co., Ltd.

**Pg 99:** Left © P & T Group; right © Huayang International Design Group

**Pg 100:** Left © Hsuyuan Kuo Architects & Associates; right © Aedas

**Pg 101:** Left © Dennis Lau & Ng Chun Man Architects & Engineers; right © Chapman Taylor

**Pg 102:** Left © 1st Image; right © Will Pryce

**Pg 103:** Left © Cathy Lee @ Arquitectonica; right © Nikken Sekkei Ltd., Photographer Hu Wenjie

**Pg 104:** Left © ECADI; right © JLL

**Pg 105:** Left © Studio City, Melco; right © Beijing Institute of Architectural Design Co., Ltd.

**Pg 106:** Left © P & T Group; right © Gensler

**Pg 107:** Left © RTKL | YiHuai Hu; right © Jiang Architects & Engineers

**Pg 110–111:** © Tim Franco, Courtesy of Kohn Pedersen Fox Associates

**Pg 112:** © Kohn Pedersen Fox Associates

**Pg 113:** © Shuhe Photo, Courtesy of Kohn Pedersen Fox Associates

**Pg 114:** © Kohn Pedersen Fox Associates

**Pg 115:** © H.G. Esch, Courtesy of Kohn Pedersen Fox Associates

**Pg 116–119:** © Robert A.M. Stern Architects

**Pg 120–123:** All © Zhubo Design Group Co., Ltd.

**Pg 124–125:** All © ECADI

**Pg 126–127:** Photos © Shui On Land; drawing © Skidmore, Owings & Merrill LLP

**Pg 130:** Top © White Swan Hotels Co., Ltd.; Bottom, Public domain

**Pg 131:** © Terri Meyer Boake

**Pg 132:** (cc-by-sa) Pubuhan

**Pg 133:** © ECADI

**Pg 134:** © Terri Meyer Boake

**Pg 135:** Top © ECADI © Daniel Safarik

**Pg 136–138:** All © Tansri Muliani

**Pg 139:** © Taipei Financial Center Corporation

**Pg 140:** Left © Daniel Safarik; right (cc-by-nd) Marc van der Chijs

**Pg 141:** Left © Joost Kuckartz; right © Anni Rao

**Pg 142:** Left (ccbysa) Nils & Araceli Jonsson; right © Dennis Lau and Ng Chun Man Archtiects and Engineers

**Pg 143:** Left (cc-by-sa) CEphoto, Uwe Aranas; right © Tansri Muliani

**Pg 144:** Left © Tansri Muliani; right © Guangzhou Design Institute

**Pg 145:** Top © PT Group; middle © Daniel Safarik; bottom © Tansri Muliani

**Pg 146:** Top and middle © Dennis Lau and Ng Chun Man Archtiects and Engineers; bottom © Georges Binder

| | |
|---|---|
| Pg 147: | Top © Beca Group; middle © Nikken Sekkei Ltd.; bottom © Tansri Muliani |
| Pg 150: | © Terri Meyer Boake |
| Pg 152–153: | All © Tongji Architectural Design (Group) Co., Ltd. |
| Pg 154: | © SOM |
| Pg 155: | Left © Skidmore, Owings & Merrill LLP; right © LV Hengzhong |
| Pg 156–157: | All © ECADI |
| Pg 158: | © Moshe Safdie Architects |
| Pg 159: | Left © Moshe Safdie Architects; top and bottom right © Arup Consulting (Shanghai) Co., Ltd. |
| Pg 160–161: | © Shanghai East China Architecture Design and Research Institute Co., Ltd. |
| Pg 162: | All © Shanghai East China Architecture Design and Research Institute Co., Ltd. |
| Pg 163: | All © Tongji Architectural Design (Group) Co., Ltd., and College of Architecture and Urban Planning of Tongji University/ ATELIER L+ |
| Pg 166–169: | © CCT Steel Co., Ltd. |
| Pg 170–171: | © Beijing Liujian Construction Group Co., Ltd. |
| Pg 172: | © Beijing Construction Engineering Group Co., Ltd. |
| Pg 173: | All © China Construction First Division Group Construction & Development Co., Ltd. |
| Pg 176: | (cc-by-sa) Verdgris |
| Pg 179–180: | © Mori; bottom (cc-by-sa) Dmottl; ECADI |
| Pg 181: | ECADI |
| Pg 182–185: | All © Daniel Safarik |

# 世界高层建筑与都市人居学会 | About the CTBUH

世界高层建筑与都市人居学会 (CTBUH) 是专注于高层建筑和未来城市的概念、设计、建设与运营的全球领先机构。学会是成立于 1969 年的非营利性组织，总部位于芝加哥的历史建筑门罗大厦，同时在上海同济大学设有亚洲办公室，意大利威尼斯建筑大学设有研究办公室，芝加哥伊利诺伊理工大学设有学术与研究办公室。学会的团队通过出版、研究、活动、工作组、网络资源和其在国际代表中广泛的网络促进全球高层建筑最新资讯的交流。学会的研究部门通过开展在可持续性和关键性发展问题上的原创性研究来引领新一代高层建筑的调查研究。学会建立了免费的高层建筑数据库——摩天大楼中心，对全球高层建筑的细节信息、图片及新闻进行每日即时更新。此外，学会还开发出测量高层建筑高度的国际标准，同时也是授予诸如"世界最高建筑"这样头衔的公认仲裁机构。

The Council on Tall Buildings and Urban Habitat (CTBUH) is the world's leading resource for professionals focused on the inception, design, construction, and operation of tall buildings and future cities. Founded in 1969 and headquartered at Chicago's historic Monroe Building, the CTBUH is a not-for-profit organization with an Asia Headquarters office at Tongji University, Shanghai; a Research Office at Iuav University, Venice, Italy; and a Chicago Research & Academic Office at the Illinois Institute of Technology. CTBUH facilitates the exchange of the latest knowledge available on tall buildings around the world through publications, research, events, working groups, web resources, and its extensive network of international representatives. The Council's research department is spearheading the investigation of the next generation of tall buildings by aiding original research on sustainability and key development issues. The Council's free database on tall buildings, The Skyscraper Center, is updated daily with detailed information, images, data, and news. The CTBUH also developed the international standards for measuring tall building height and is recognized as the arbiter for bestowing such designations as "The World's Tallest Building."

# 中国高层建筑国际交流委员会 | About the CITAB

中国高层建筑国际交流委员会（CITAB）成立于2013年，是设于中国建筑学会的学术组织，旨在加强我国高层建筑规划、设计、施工、材料、设备、管理、维护等单位和专业人员与国外相关组织与专家的交流，促进我国高层建筑的高水平发展，提升行业单位在高层建筑领域的国际影响。其主要任务包括：代表中国建筑学会与高层建筑相关国际组织合作，组织和参与相关学术活动；组织中国高层建筑领域的国际学术活动和优秀高层建筑的评选活动，引导高层建筑健康发展；收集中国高层建筑数据，了解世界高层建筑发展动态，提供世界高层建筑的前沿讯息，为科学研究提供导向。

The China International Exchange Committee for Tall Buildings (CITAB) is one of academic organizations affiliated to Architecture Society of China. Founded in 2013, CITAB is aiming to motivate international exchange of urban planning, tall building design, construction, equipments and facilities management organizations involved in this industry, to enhance various organizations international influence via level up tall building development in China. CITAB is currently facilitating the exchange and research as the following three aspects: to network academic exchange cooperating with international organizations; to organize China Tall Building Awards program, leading tall building development benchmark; to investigate and introduce China and global tall building information and knowledge for industry development and research.

# CTBUH Organizational Structure & Members (as of March 2016)
# 世界高层建筑与都市人居学会组织构架和会员（截至2016年3月）

## Board of Trustees
**Chairman:** David Malott, *Kohn Pedersen Fox*, USA
**Vice-Chairman:** Timothy Johnson, *NBBJ*, USA
**Executive Director:** Antony Wood, *CTBUH / Illinois Institute of Technology*, USA / *Tongji University*, China
**Treasurer:** Steve Watts, *Alinea Consulting LLP*, UK
**Secretary:** Tim Neal, *Arcadis*, UK
**Trustee:** Mounib Hammoud, *Jeddah Economic Company*, Saudi Arabia
**Trustee:** Dennis Poon, *Thornton Tomasetti*, USA
**Trustee:** Abrar Sheriff, *Turner Construction*, USA
**Trustee:** Kam-Chuen (Vincent) Tse, *Parsons Brinckerhoff*, Hong Kong

## China Office Board
Murilo Bonilha, *United Technologies Research Center*, Shanghai
Jianping Gu, *Shanghai Tower Construction & Development*, Shanghai
Eric Lee, *JLL*, Hong Kong
David Malott, *Kohn Pedersen Fox*, New York, USA
Wai Ming Tsang, *Ping An Real Estate*, Shenzhen
Antony Wood, *CTBUH / Illinois Institute of Technology*, USA / *Tongji University*, China
Changfu Wu, *Tongji University*, Shanghai
Junjie Zhang, *ECADI*, Shanghai
Kui Zhuang, *CCDI*, Shanghai

## Staff / Contributors
**Executive Director:** Antony Wood
**Associate Director:** Steven Henry
**China Office Director:** Daniel Safarik
**Associate China Office Director:** Xia Sun
**Operations Manager:** Patti Thurmond
**Research Manager:** Dario Trabucco
**Leader Coordinator / Events Manager:** Jessica Rinkel
**Digital Platforms Manager:** Son Dang
**Production Manager:** Marty Carver
**Membership Coordinator:** Carissa Devereux
**Communications Manager:** Jason Gabel
**Staff Writer/ Media Associate:** Benjamin Mandel
**Website Content Editor:** Aric Austermann
**Production Associate:** Kristen Dobbins
**Events Assistant:** Chuck Thiel
**Research Assistant / China Operations:** Peng Du
**Skyscraper Database Editor:** Marshall Gerometta
**Skyscraper Database Assistant:** Will Miranda
**Publications Associate:** Tansri Muliani
**General Counsel:** Matt Rossetti
**Special Media Correspondent:** Chris Bentley

## Advisory Group
Ahmad K. Abdelrazaq, *Samsung Corporation*, Korea
Jim Bilger, *CBRE*, USA
Joseph Chou, *Taipei Financial Center Corporation*, Taiwan
Christopher Colasanti, *JBB*, USA
Donald Davies, *Magnusson Klemencic*, USA
Scott Duncan, *SOM*, USA
John Gaskin, *Brookfield Multiplex*, Australia
Jean-Claude Gerardy, *ArcelorMittal*, Luxembourg
Faudziah Ibrahim, *KLCC*, Malaysia
Abdo Kardous, *Hill International*, USA
Stephen Lai, *Rider Levett Bucknall*, China
Anand Pandit, *Lotus Group*, India
James Parakh, *City of Toronto*, Canada
Mic Patterson, *Enclos*, USA
Glen Pederick, *Waterman International*, Australia
Robert Pratt, *Tishman Speyer Properties*, China
Peter Weismantle, *Adrian Smith + Gordon Gill Architecture*, USA

## Working Group Co-Chairs
**BIM:** Stuart Bull
**Demolition:** Dario Trabucco
**Façade Access:** Lance McMasters, Kevin Thompson & Peter Weismantle
**High Performance Façades:** Christopher Drew & Mikkel Kragh
**Legal Aspects of Tall Buildings:** Victor Madeira Filho & Arthur Wellington
**Performance Based Seismic Design:** Ron Klemencic & John Viise
**Security:** Sean Ahrens & Caroline Field
**Sustainable Design:** Antony Wood
**Tall Timber:** Carsten Hein & Volker Schmid

## Committee Chairs
**Urban Habitat / Urban Design:** James Parakh, *City of Toronto Planning Department*, Canada
**Expert Peer Review Committee:** Antony Wood, *CTBUH / Illinois Institute of Technology*, USA / *Tongji University*, China
**Height & Data:** Peter Weismantle, *Adrian Smith + Gordon Gill Architecture*, USA
**Awards:** Mun Summ Wong, *WOHA*, Singapore
**Expert Chinese Translation Committee:** Nengjun Luo, *CITIC HEYE Investment CO., LTD.*, China
**Skyscraper Center Editorial Board:** Marshall Gerometta, *CTBUH*, USA
**Young Professionals:** Sasha Zeljic, *Gensler*, USA

## Regional Representatives
**Australia:** Bruce Wolfe, *Conrad Gargett Architecture*
**Belgium:** Georges Binder, *Buildings & Data S.A.*
**Brazil:** Antonio Macedo, *EcoBuilding Consultoria*
**Cambodia:** Michel Cassagnes, *Archetype Group*
**Canada:** Richard Witt, *Quadrangle Architects*
**China:** Daniel Safarik, *CTBUH*
**Costa Rica:** Ronald Steinvorth, *IECA International*
**Finland:** Santeri Suoranta, *KONE Industrial, Ltd.*
**France:** Trino Beltran, *Bouygues Construction*
**Germany:** Roland Bechmann, *Werner Sobek Stuttgart GmbH*
**Greece:** Alexios Vandoros, *VaCo Group*
**India:** Girish Dravid, *Sterling Engineering*
**Indonesia:** Tiyok Prasetyoadi, *PDW Architects*
**Iran:** Matin Alaghmandan, *University of Tehran*
**Israel:** Israel David, *David Engineers*
**Italy:** Dario Trabucco, *Iuav University of Venice*
**Lebanon:** Ramy El-Khoury, *Rafik El-Khoury & Partners*
**Malaysia:** Matthew Gaal, *Cox Architecture*
**Mongolia:** Tony Mills, *Archetype Group*
**Myanmar:** Mark Petrovic, *Archetype Group*
**Nigeria:** Shola Sanni, *Sanni, Ojo & Partners Consulting*
**Philippines:** Felino A. Palafox Jr., *Palafox Associates*
**Poland:** Ryszard M. Kowalczyk, *University of Beira Interior*
**Qatar:** Shaukat Ali, *KEO International Consultants*
**Russia:** Elena A. Shuvalova, *Lobby Agency*
**Saudi Arabia:** Bassam Al-Bassam, *Rayadah Investment Company, KSA*
**Scandinavia:** Julian Chen, *Henning Larsen Architects*
**South Korea:** Dr. Kwang Ryang Chung, *Dongyang Structural Engineers Co., Ltd*
**Spain:** Iñigo Ortiz Diez de Tortosa, *Ortiz Leon Arquitectos*
**Sri Lanka:** Shiromal Fernando, *Civil and Structural Engineering Consultants (Pvt.) Ltd*
**UAE:** Dean McGrail, *WSP Middle East*
**United Kingdom:** Steve Watts, *alinea consulting LLP*
**Vietnam:** Phan Quang Minh, *National University of Civil Engineering*

## CTBUH Organizational Members
*(as of March 2016) http://membership.ctbuh.org*

### Supporting Contributors 顶级会员
Arcadis
ARK Reza Kabul Architects
Beijing Fortune Lighting System Engineering Co., Ltd.
BuroHappold Engineering
CCDI Group 悉地国际
CITIC HEYE Investment CO., LTD. 中信和业投资有限公司
D2E International VT Consultants Ltd.
Dow Corning Corporation
Emaar Properties
HSB Sundsfastigheter
Hudson Yards
Illinois Institute of Technology
IUAV University of Venice
Jeddah Economic Company
Kingdom Real Estate Development
Kohn Pedersen Fox Associates
KONE Industrial
Lotte Engineering & Construction
National Engineering Bureau
New World Development Company Limited 新世界发展有限公司
Otis Elevator Company
Pace Development Corporation Plc.
Ping An Financial Center Construction & Development 平安金融中心建设发展有限公司
Property Markets Group
Samsung C&T Corporation
Schindler Top Range Division
Shanghai Tower Construction & Development 上海中心大厦建设发展有限公司
Shenzhen Aube Architectural Engineering Design
Shenzhen Parkland Real Estate Development Co., Ltd.
Skidmore, Owings & Merrill
Sun Hung Kai Properties Limited 新鸿基地产
Taipei Financial Center Corp. 台北金融大楼公司
Tongji University 同济大学
Turner Construction Company
Underwriters Laboratories
Wentworth House Partnership Limited
WSP | Parsons Brinckerhoff
Zhongtian Urban Development Group

### Patrons 赞助会员
Blume Foundation
BMT Fluid Mechanics
Citic Pacific 中信泰富
DeSimone Consulting Engineers
Durst Organization, The
East China Architectural Design & Research Institute 华东建筑设计研究总院
Empire State Realty Trust
Fly Service Engineering
Forest City Ratner Companies
Gensler
Hoboken Brownstone
HOK, Inc.
Hongkong Land 香港置地有限公司
ISA Architecture 上海建筑设计研究院有限公司
KLCC Property Holdings Berhad
Kuraray America, Inc.
Langan
Meinhardt Group International
NBBJ
Permasteelisa Group
Ridley
Rowan Williams Davies & Irwin
Shenzhen Overseas Chinese Town 深圳华侨城房地产有限公司
SL Green Management
Studio Libeskind
Thornton Tomasetti
Thyssenkrupp Elevator
Tishman Speyer
United Technologies Corporation
Wirth Research
Zuhair Fayez Partnership

### Donors 高级会员
A&H Tuned Mass Dampers
Adrian Smith + Gordon Gill Architecture
ALT Limited
American Institute of Steel Construction
Aon Fire Protection Engineering
Arcadis Australia Pacific
Arquitectonica International
Arup
Aurecon
BALA Engineers
Broad Sustainable Building Co. 远大集团
Brookfield Multiplex
CBRE Group
Enclos Corp.
Fender Katsalidis

Frasers Property
Guangzhou Yuexiu City Construction Jones Lang LaSalle Property Management Co., Ltd.
Halfen United States
Henning Larsen Architects
Hill International
Hilti
Jensen Hughes
JORDAHL
Jotun Group, The
Laing O'Rourke
Larsen & Toubro
Leslie E. Robertson Associates
Magnusson Klemencic Associates
MAKE
McNamara / Salvia, Inc.
Nishkian Menninger Consulting and Structural Engineers
Outokumpu
PDW Architects
Pei Cobb Freed & Partners
Pelli Clarke Pelli Architects
Pickard Chilton Architects
Plaza Construction
PLP Architecture
PNB Merdeka Ventures Sdn. Berhad
PT Gistama Intisemesta
Quadrangle Architects
SAMOO Architects and Engineers
Saudi Binladin Group / ABC Division
Schüco
Severud Associates Consulting Engineers
Shanghai Construction (Group) General 上海建工（集团）总公司
SHoP Architects
Sika Services AG
Sinar Mas Group - APP China 金光集团—APP（中国）
Solomon Cordwell Buenz
Spiritos Properties LLC
Studio Gang Architects
Syska Hennessy Group
TAV Construction
Terracon
Time Equities
Tongji Architectural Design Group 同济大学建筑设计研究院（集团）有限公司
UltraTech Cement Sri-Lanka
Walsh Construction Company
Walter P. Moore and Associates
WATG
Werner Voss + Partner
William Hare
Wordsearch
Zaha Hadid Architects

## Contributors 中级会员

Aedas 凯达环球
AkzoNobel
Alimak Hek
Alinea Consulting
Allford Hall Monaghan Morris
Altitude Facade Access Consulting
Alvine Engineering
AMSYSCO
Andrew Lee King Fun & Associates Architects Ltd.
ArcelorMittal
architectsAlliance
Architectural Design & Research Institute of South China University of Technology
Architectural Design & Research Institute of Tsinghua University 清华大学建筑设计研究院
Architectus
Barker Mohandas, LLC
Bates Smart
Benson Industries Inc.
BG&E
bKL Architecture
Bonacci Group
Bosa Properties Inc.
Boundary Layer Wind Tunnel Laboratory
Bouygues Batiment International
British Land Company

Broadway Malyan
Brookfield Property Group
Brunkeberg Systems
Cadillac Fairview
Canary Wharf Group
Canderel Management
CB Engineers
CCL
Cerami & Associates
Cermak Peterka Petersen
Chapman Taylor
China Electronics Engineering Design Institute
Clark Construction
Code Consultants, Inc.
Conrad Gargett
Continental Automated Buildings Association
Cosentini Associates
Cottee Parker Architects
CoxGomyl
CS Group Construction Specialties Company
CS Structural Engineering
CTSR Properties
Cubic Architects
Dar Al-Handasah (Shair & Partners)
Davy Sukamta & Partners Structural Engineers
DB Realty Ltd.
DCA Architects
DCI Engineers
DDG
Deerns
DIALOG
Dong Yang Structural Engineers
dwp|suters
Edwards and Zuck Consulting Engineers
Elenberg Fraser
Elevating Studio Pte. Ltd.
EllisDon Corporation
Euclid Chemical Company, The
Eversendai Engineering Qatar
Facade Tectonics
Farrells
Foster + Partners
FXFOWLE Architects
GEI Consultants
GERB Vibration Control Systems (USA/Germany)
GGLO
Global Wind Technology Services
Glumac
gmp • Architekten von Gerkan, Marg und Partner GbR
Goettsch Partners
Grace Construction Products
Gradient Wind Engineering Inc.
Graziani + Corazza Architects
Guangzhou Design Institute 广州市设计院
Halvorson and Partners: A WSP | Parsons Brinckerhoff Company
Hariri Pontarini Architects
Harman Group, The
Hathaway Dinwiddie
Heller Manus Architects
HKA Elevator Consulting
Housing and Development Board
Humphrey & Partners Architects, L.P.
Hutchinson Builders
Hysan Development Company Limited
IDOM UK Ltd.
Irwinconsult Pty.
Israeli Association of Construction and Infrastructure Engineers
ITT Enidine
JAHN
Jangho Group
Jaros, Baum & Bolles
JDS Development Group
Jiang Architects & Engineers
JLL
John Portman & Associates
Kajima Design
KEO International Consultants
KHP Konig und Heunisch Planungsgesellschaft
Kinemetrics Inc.

Langdon & Seah Singapore
LeMessurier
Lend Lease
Lusail Real Estate Development Company
M Moser Associates 穆氏有限公司
Maeda Corporation
MAURER AG
MicroShade A/S
Mori Building Company 森大厦有限公司
Nabih Youssef & Associates
National Fire Protection Association
National Institute of Standards and Technology
NIKKEN SEKKEI LTD
Norman Disney & Young
OMA
Omrania & Associates
Ornamental Metal Institute of New York
Pakubuwono Development, The
Palafox Associates
Pappageorge Haymes Partners
Pavarini McGovern
Pepper Construction Company
Perkins + Will
Plus Architecture
Probuild
Profica
Project Planning and Management Pty Ltd
R.G. Vanderweil Engineers
Ramboll
RAW Design
Read Jones Christoffersen
Related Midwest
Rhode Partners
Richard Meier & Partners
RMC International
Ronald Lu & Partners 吕元祥建筑师事务所
Royal HaskoningDHV
Sanni, Ojo & Partners
Sematic Elevator Products
Shimizu Corporation
SilverEdge Systems Software
Silverstein Properties
Skanska
SkyriseCities
Soyak Construction and Trading Co.
Spectrum Metal Finishing Inc.
Stanley D. Lindsey & Associates
Stauch Vorster Architects
Steel Institute of New York
Stein Ltd.
SuperTEC (Super-Tall Building Design & Engineering Tech Research Center)
Surface Design
SWA Group
Taisei Corporation
Takenaka Corporation
Tate Access Floors
Taylor Devices, Inc.
TMG Partners
Trimble Solutions Corporation
TSNIIEP for Residential and Public Buildings
Uniestate
University of Illinois at Urbana–Champaign
Vetrocare
Waterman AHW (Vic) Pty Ltd.
Weischede, Herrmann und Partners
Werner Sobek Group
Wilkinson Eyre Architects
WOHA Architects
Woods Bagot
WTM Engineers International
WZMH Architects
Y. A. Yashar Architects

## Participants 普通会员

There are an additional 279 members of the Council at the Participant level. Please see online for the full member list: http://members.ctbuh.org

---

Supporting Contributors are those who contribute $10,000; Patrons: $6,000; Donors: $3,000; Contributors: $1,500; Participants: $750.
年度会费：顶级会员$10 000；赞助会员$6 000；高级会员$3 000；中级会员$1 500；普通会员$750。

# 中国高层建筑国际交流委员会组织架构和会员

## 领导小组

### 主任委员
张俊杰，华东建筑设计研究总院，院长

### 副主任委员
李国强，同济大学，教授
丁洁民，同济大学建筑设计研究院（集团）有限公司，总裁、总工程师
刘恩芳，上海建筑设计研究院有限公司，董事长
刘彦生，清华大学建筑设计研究院有限公司，总工程师
梁金桐，奥雅纳工程咨询（上海）有限公司，董事
戴立先，中建钢构有限公司，总工程师
桂学文，中南建筑设计院股份有限公司，总建筑师
曲大为，北京建工集团有限责任公司，副总工程师
黄惠菁，广州市设计院，总建筑师
朱毅，宋腾添玛沙帝建筑工程设计咨询（上海）有限公司，资深副总裁
贾朝晖，绿地控股集团，副总建筑师

### 团体委员
CCDI 悉地国际
AECOM
华南理工大学建筑设计研究院
华南理工大学土木与交通学院
天津大学建筑工程学院
GENSLER
杭萧钢构股份有限公司
清华大学建筑学院
上海三菱电梯有限公司
浙江省建筑设计研究院
中国建筑东北设计研究院有限公司
利沛建筑技术咨询（上海）有限公司

### 个人委员
周建龙
肖从真
韩林海
龚 剑
范 重
傅学怡
刘 鹏
蒋欢军
马 泷
贾朝晖
戴立先

## 专业委员会

CITAB 下设 10 个专业委员会，负责组织和高层建筑相关的各个专业领域的学术研讨和交流，组织国内外的技术参观、考察活动，编制技术要点、指导书、手册、指南等。

### 建筑
主任委员单位：华东建筑设计研究总院
副主任委员单位：同济大学建筑设计研究院（集团）有限公司
上海建筑设计研究院有限公司
清华大学建筑设计研究院有限公司
广州市设计院

### 规划
主任委员单位：上海建筑设计研究院有限公司
副主任委员单位：同济大学建筑设计研究院（集团）有限公司
华东建筑设计研究总院
奥雅纳工程咨询（上海）有限公司
SWECO
广州市设计院
北京建工集团有限责任公司

### 结构
主任委员单位：同济大学建筑设计研究院（集团）有限公司
副主任委员单位：同济大学
华东建筑设计研究总院
宋腾添玛沙帝建筑工程设计咨询（上海）有限公司（Thornton Tomasetti）
奥雅纳工程咨询（上海）有限公司
清华大学建筑设计研究院有限公司
广州市设计院

### 建造
主任委员单位：中建钢构有限公司
副主任委员单位：北京建工集团有限责任公司

### 机电
主任委员单位：科进/柏诚集团（中国）（WSP | Parsons Brinckerhoff, China）
副主任委员单位：华东建筑设计研究总院
同济大学建筑设计研究院（集团）有限公司
奥雅纳工程咨询（上海）有限公司
广州市设计院
北京建工集团有限责任公司

### 安全与防灾
主任委员单位：奥雅纳工程咨询（上海）有限公司
副主任委员单位：华东建筑设计研究总院

### BIM
主任委员单位：华东建筑设计研究总院
副主任委员单位：北京建工集团有限责任公司
清华大学建筑设计研究院有限公司
奥雅纳工程咨询（上海）有限公司

### 运营和管理
主任委员单位：绿地控股集团
副主任委员单位：奥雅纳工程咨询（上海）有限公司
宋腾添玛沙帝建筑工程设计咨询（上海）有限公司（Thornton Tomasetti）
华东建筑设计研究总院
广州市设计院

### 绿色节能
主任委员单位：清华大学建筑设计研究院有限公司
副主任委员单位：华东建筑设计研究总院
北京建工集团有限责任公司
奥雅纳工程咨询（上海）有限公司
广州市设计院

### 材料和幕墙
副主任委员单位：奥雅纳工程咨询（上海）有限公司

## 主要工作内容

CITAB 的工作围绕着评价中心、信息中心和交流中心这"三个中心"展开，具体如下：

评价中心：组织高层建筑领域的国际学术活动和优秀高层建筑的评选和表彰活动，引导高层建筑健康发展；
信息中心：负责收集中国高层建筑数据，了解世界高层建筑发展动态，提供世界高层建筑的前沿讯息，为科学研究提供导向；
交流中心：组织国内外学术和技术活动的交流，引进国外的先进理念和技术，提升我国在高层建筑领域的国际影响。

## 以往主要活动一览

2014 年 7 月 17 日，上海中心的技术参观活动
2014 年 7 月 18 日，第一届中韩高层建筑论坛（中国上海）
2014 年 7 月 31 日，高层建筑结构设计优化技术研讨会
2014 年 9 月 18 日，第一届中美高层建筑研讨会——建筑与结构设计的创新与整体协同（中国上海）
2014 年 9 月 25 日，高柔建筑的黏弹性耦合阻尼器减振作用研讨会
2014 年 10 月 27 日，复杂超限高层建筑混合结构设计与案例讲座
2015 年 4 月 21 日，高层建筑项目运营研讨会
2015 年 5 月 27 日，超高层钢结构施工技术论坛
2015 年 6 月 30 日，2015 年上海城市综合体安全高峰论坛——城市综合体的消防与安防问题
2015 年 7 月 6 日，CITAB 系列讲座："基于限额设计的超高层建筑绿色设计策略：以北京 CBD Z1B 项目为例"
2015 年 7 月 17 日，超限高层建筑审查要点若干问题及工程案例学术报告会
2015 年 9 月 10 日，高层建筑结构黏滞阻尼器应用技术报告会
2015 年 10 月 15~16 日，第二届中日韩高层建筑论坛（韩国首尔）
2015 年 10 月，首届 CITAB-CTBUH 中国高层建筑奖项评选活动正式发布
2015 年 11 月 3 日，超高层建筑夜景照明设计学术讲座
2015 年 11 月 4 日，第二届中美高层建筑研讨会（中国上海）
2015 年 12 月 10 日，"以区域的视角——高层建筑群体设计与多地块联合开发模式下的建筑设计思考"论坛
2015 年 12 月 11 日，超高层及超高构物结构工程建造关键技术发展学术报告会
2016 年 1 月 21 日，首届 CITAB-CTBUH 中国高层建筑奖项评选评委会议召开（中国上海）
2016 年 4 月 21 日，第三届中美高层建筑研讨会（美国旧金山）
2016 年 5 月 13 日，首届 CITAB-CTBUH 中国高层建筑奖获奖作品研讨会暨颁奖典礼召开（中国上海）

更多信息，请访问网站：www.citab.org

年度会费：主任委员单位 50 000 元；副主任委员单位 20 000 元；团体委员 5 000 元；个人委员 100 元

# CITAB Organizational Structure & Members

## Leading Group

### Chairman
Junjie Zhang, Director, Shanghai East China Architecture Design and Research Institute Co., Ltd.

### Cochairman
Guoqiang Li, Professor, Tongji University
Jiemin Ding, President and Chief Engineer, Tongji Architectural Design (Group) Co., Ltd.
Enfang Liu, President, Institute of Shanghai Architectural Design and Research Co., Ltd
Yansheng Liu, Chief Engineer, Architectural Design & Research Institute of Tsinghua University
Jintong Liang, Director, Arup Shanghai
Lixian Dai, Chief Engineer, China Construction Steel Structure Co., Ltd.
Xuewen Gui, Chief Architect, Central-south Architectural Design Institute Co., Ltd.
Dawei Qu, Deputy Chief Engineer, Beijing Construction Engineering Group
Huijing Huang, Deputy Chief Architect, Guangzhou Design Institute
Yi Zhu, Senior Vice President, Thornton Tomasetti Shanghai
Zhaohui Jia, Deputy Chief Architect, Greenland Group

### Organization Member
CCDI
AECOM
Architectural Design & Research Institute of SCUT
School of Civil Engineering and Transportation of South China University of Technology
School of Civil Engineering of Tianjin University
Gensler
Hangxiao Steel Structure Co. Ltd.
School of Architecture of Tsinghua University
Shanghai Mitsubishi Elevator Co., Ltd.
Zhejiang Province Institute of Architectural Design and Research
China Northeast Architectural Design & Research Institute Co., Ltd.
Lerch Bates (China) Limited, Shanghai

### Individual Member
Jianlong Zhou
Congzhen Xiao
Linhai Han
Jian Gong
Zhong Fan
Xueyi Fu
Peng Liu
Huanjun Jiang
Long Ma
Zhaohui Jia
Lixian Dai

## Professional Committee
CITAB has ten professional committees responsible for a variety of activities, including hosting seminars and academic exchanges on disciplines related to tall buildings, organizing technical visits, developing technical guidelines and handbooks, etc.

### Architecture
Chairman Organization:
Shanghai East China Architecture Design and Research Institute Co., Ltd.
Deputy Chairman Organization: Tongji Architectural Design (Group) Co., Ltd.
Institute of Shanghai Architectural Design and Research Co., Ltd
Architectural Design & Research Institute of Tsinghua University
Guangzhou Design Institute

### Planning
Chairman Organization: Institute of Shanghai Architectural Design and Research Co., Ltd
Deputy Chairman Organization: Tongji Architectural Design (Group) Co., Ltd.
Shanghai East China Architecture Design and Research Institute Co., Ltd.
Arup Shanghai
SWECO
Guangzhou Design Institute
Beijing Construction Engineering Group

### Structure
Chairman Organization: Tongji Architectural Design (Group) Co., Ltd.
Deputy Chairman Organization: Tongji University
Shanghai East China Architecture Design and Research Institute Co., Ltd.
Thornton Tomasetti Shanghai
Arup Shanghai
Architectural Design & Research Institute of Tsinghua University
Guangzhou Design Institute

### Construction
Chairman Organization: China Construction Steel Structure Co., Ltd.
Deputy Chairman Organization: Beijing Construction Engineering Group

### MEP
Chairman Organization: WSP | Parsons Brinckerhoff, China
Deputy Chairman Organization: Shanghai East China Architecture Design and Research Institute Co., Ltd.
Tongji Architectural Design (Group) Co., Ltd.
Arup Shanghai
Guangzhou Design Institute
Beijing Construction Engineering Group

### Safety
Chairman Organization: Arup Shanghai
Deputy Chairman Organization: Shanghai East China Architecture Design and Research Institute Co., Ltd.

### BIM
Chairman Organization: Shanghai East China Architecture Design and Research Institute Co., Ltd.
Deputy Chairman Organization: Beijing Construction Engineering Group
Architectural Design & Research Institute of Tsinghua University
Arup Shanghai

### Operation and Management
Chairman Organization: Greenland Group
Deputy Chairman Organization: Arup Shanghai
Thornton Tomasetti Shanghai
Shanghai East China Architecture Design and Research Institute Co., Ltd.
Guangzhou Design Institute

### Environment
Chairman Organization: Architectural Design & Research Institute of Tsinghua University
Deputy Chairman Organization: Shanghai East China Architecture Design and Research Institute Co., Ltd.
Beijing Construction Engineering Group
Arup Shanghai
Guangzhou Design Institute

### Material and Curtain Wall
Deputy Chairman Organization: Arup Shanghai

## Scope
CITAB functions as an evaluation, information and communication center which performs the following activities:
Evaluation: organize international academic activities on tall building and put forward awards praising excellent tall buildings, in order to facilitate the development of this area.
Information: collect information on tall buildings in China, track the trends and progress all over the world, and publish news about international tall buildings, so as to guide related research.
Communication: organize academic and technical exchanges at home and abroad, introduce advance concepts and technologies, and improve the international fame of tall buildings in China.

## Activities

17 July, 2014 Technical visits at Shanghai Center
18 July, 2014 The First Sino-Korean Tall Building Forum (Shanghai, China)
31 July, 2014 Seminar: Tall Building Structure Optimization
18 Sep, 2014 The First Sino-American Tall Building Seminar: Innovation and Coordination in Architectural and Structural Design (Shanghai, China)
25 Sep, 2014 Seminar: The Damping Effect of Viscoelastic Coupling Damper of High Flexible Buildings
27 Oct, 2014 Lecture: Designs and Cases of Complex Ultra-limit Mixed Structures in Tall Buildings
21 Apr, 2015 Seminar: Project Operation of Tall Buildings
27 May, 2015 Ultra-tall Steel Structure Construction Technical Forum
30 June, 2015 Safety of Shanghai Mixed-use Projects Summit 2015: Fire Control and Security Guard of Urban Mixed-use Projects
6, July, 2015 Lectures: Green Design Strategy of Ultra-tall Buildings based on Quota Design with a Case Study on Beijing CBD Z1B
17 July, 2015 Symposium: Issues and Engineering Cases of Ultra-limit Tall Building Auditing
10 Sep, 2015 Symposium: Applied Technology of Fluid Viscous Damper in Tall Building Structure
15-16 Oct, 2015 The Second Sino-Korean Tall Building Forum (Seoul, Korea)
Oct, 2015 The First CITAB-CTBUH China Tall Building Awards campaign is officially started
3 Nov, 2015 Lecture: Night Illumination of Ultra-tall Buildings
4 Nov, 2015 The Second Sino-American Tall Building Seminar (Shanghai, China)
10 Dec, 2015 Forum: Regional Perspective: Group Design of Tall Buildings and Architectural Design in the Mode of Multi-sites Collateral Development
11 Dec, 2015 Symposium: Key Technical Advances Related to Ultra-tall Buildings and Structures Engineering
21 Jan, 2016 The Appraisal Meeting of the First CITAB-CTBUH China Tall Building Awards (Shanghai, China)
21 Apr, 2016 The Third Sino-American Tall Building Seminar (San Francisco, USA)
13 May, 2016 Award Ceremony and Seminars of the First CITAB-CTBUH China Tall Building Awards (Shanghai, China)

For more information, please visit: *www.citab.org*

Annual Fee: Chairman Organization ¥50,000; Deputy Chairman Organization ¥20,000; Organization Member ¥5,000; Individual Member ¥100.